WORKBOOK FOR

Introductory Statistics for the Behavioral Sciences

Alternate Third Edition with Computer Software

Robert B. Ewen

Harcourt Brace Jovanovich, Publishers

San Diego New York Chicago Austin Washington, D.C.
London Sydney Tokyo Toronto

to two nice people————
Jack Cohen and Joan Welkowitz

COVER ART BY EDWARD A. BURKE

ISBN: 0-15-545985-6

Printed in the United States of America

Contents

Introduction

Many behavioral science students approach the study of statistics with fear and trepidation, due perhaps to previous unpleasant experiences with mathematical concepts. Nevertheless, statistics is extremely important for those majoring in the behavioral and social sciences. You may design research of your own in the future, and you will surely have to read and critically evaluate experiments and statistical procedures carried out by others. Yet statistics need not be unduly difficult to learn, provided that you take matters one step at a time and make sure that you understand each topic before proceeding to the next.

In particular, step-by-step homework practice is essential to learning and understanding this material. To be sure, the development of hand calculators and personal computers has eliminated the need for detailed (and often tedious) calculations: after entering the data, you simply press a few keys, and the machine promptly reports the standard deviation, or mean, or correlation coefficient, or whatever. Excellent software is now available, such as MYSTAT, which will perform most of the procedures used in this book. However, the computer normally does not show how the calculations were performed. Thus it is highly desirable to go behind the scenes and examine the computational procedures firsthand, so that you can see how the formulas work and what they are designed to accomplish. The purpose of this workbook is to help you obtain the necessary practice.

The following objectives have guided the development of this workbook:

1. The workbook should be interesting. No one can make a statistics workbook as entertaining as a popular novel, but a dry and lifeless approach can only lead to boredom and distaste for statistics and should be avoided insofar as is possible.

2. Numerical computations should be kept within reasonable bounds. Burdening introductory statistics students with hours of laborious mathematical calculations can hardly be considered a meaningful educational experience. All too often, students lose sight of the general procedures and underlying rationales when faced with endless masses of computations. Also, any enthusiasm the student may have had for statistics is likely to be lost in the process. Computations are necessary to develop understanding of the material, but there is no reason why the computations cannot be relatively simple and involve a relatively small number of figures.

Therefore, it is important for you to understand that many numerical examples in this workbook are designed with the goal of ready computation in mind. On page 1, you will see some hypothetical data that will be used in many of the homework problems. In actual practice, it would not be a good idea to draw small samples of only five or ten cases, as is done with the "University C" and "University D" data. Similarly, other problems in this workbook often involve a smaller sample than would be desirable in a real situation. If you can deal correctly with the problems in this workbook, you will have no difficulty with real problems that involve exactly the same statistical operations but have more data to be analyzed.

3. The workbook should stress understanding as well as computation. Carrying out computations correctly is highly commendable, but leaves something to be desired if you do not understand the reasons for what you are doing. Therefore, this workbook includes questions that are designed to help you evaluate your understanding of the various procedures, as well as questions designed to give you practice in performing the necessary calculations.

4. There should be some continuity across different units. If a workbook consists of a potpourri of problems, where those in one chapter have nothing to do with those in the next, it is difficult for students to appreciate the relationships among different areas. Therefore, the hypothetical "University" data are provided in order to establish some continuity across the various topics covered in this workbook. These data will not be suitable for all problems, so new data will be introduced at various points. However, the "University" data should allow you to get a fairly good idea of how the researcher goes from a mass of numbers to various descriptive statistics, the purpose of which is to make the data more understandable, and to various inferential statistics, the purpose of which is to permit the researcher to test scientific hypotheses.

5. The workbook should highlight formulas and key ideas for ready reference. At the beginning of each chapter, you will find a "reminder" to provide a ready reference to the formulas and important ideas for that chapter. This workbook is not intended to stand by itself, and you should use the reminders for review and reference only and use your textbook to learn the material. In a few cases where the computation of a statistic is moderately complicated, a numerical example has been included to further clarify the procedures involved.

6. The workbook should deal with the choice of the correct procedure. All too often, writers in the area of statistics ignore the fact that the selection of the appropriate statistic is often as much of a problem to the beginning student (and to the more experienced one as well!) as is the proper computation of that statistic. Therefore, exercises are included which deal with *which* procedure to use in a given situation.

Most chapters in this workbook consist of three sections: (1) *Reminder:* A summary of important formulas and concepts. The "Reminder" sections are also incorporated into the textbook as chapter summaries, so that you will have this information available for ready reference after the homework has been handed in. (2) *Problems:* Basic calculational problems, and thought problems designed to test your understanding of statistical concepts and issues. (3) *Additional Computational Practice:* Problems that provide more extensive computational practice. At the discretion of your instructor, these problems may either be assigned to all students or just to those who have had difficulty with the preceding section.

Note that answers to selected problems have been included at the end of the workbook. This will enable you to check your work in at least some instances and gain immediate feedback. However, answers to some problems have not been provided. The reason for this is to simulate realistic conditions: when you perform a research study and analyze the data, there will be no back of the book to which you can turn in order to find out if your calculations were correct.

The behavioral or social scientist with a competent understanding of statistics has at hand a collection of quality tools, which when applied correctly will aid in the development of proper data analysis. I hope that you will develop an appreciation and liking for the techniques of statistics, and that they will serve you well in the years to come.

Glossary of Symbols

Numbers in parentheses indicate the chapter in which the symbol first appears.

a_{YX}	*Y* intercept of linear regression line for predicting *Y* from *X* (12)
α	criterion (or level) of significance; probability of Type I error (10)
b_{YX}	slope of linear regression line for predicting *Y* from *X* (12)
β	probability of Type II error (10)
$1 - \beta$	power (14)
cf	cumulative frequency (2)
χ^2	chi square (17)
D	difference between two scores or ranks (11)
$\bar{D}$	mean of the *D*s (11)
df	degrees of freedom (10)
df_B	degrees of freedom between groups (15)
df_W	degrees of freedom within groups (15)
df_1	degrees of freedom for factor 1 (16)
df_2	degrees of freedom for factor 2 (16)
$df_{1 \times 2}$	degrees of freedom for interaction (16)
δ	delta (14)
ε	epsilon (15)
ε_R	epsilon applied to ranks (18)
f	frequency (2)
f_e	expected frequency (17)
f_m	number of negative difference scores (18)
f_o	observed frequency (17)
f_p	number of positive difference scores (18)
F	statistic following the *F* distribution (15)
$G\ Mdn$	grand median (18)
γ	effect size, gamma (14)
h	interval size (3)
H	Kruskal–Wallis *H* (18)
$H\%$	percent of subjects in all intervals higher than the critical one (3)
H_0	null hypothesis (10)
H_1	alternative hypothesis (10)

i	case number (1)
$I\%$	percent of subjects in the critical interval (3)
k	a constant (1)
k	number of groups (or the last group) (15)
$L\%$	percent of subjects in all intervals below the critical one (3)
LRL	lower real limit (3)
LSD	Fisher protected t test (15)
Mdn	median (4)
MS	mean square (15)
MS_B	mean square between groups (15)
MS_W	mean square within groups (15)
MS_1	mean square for factor 1 (16)
MS_2	mean square for factor 2 (16)
$MS_{1\times 2}$	mean square for interaction (16)
μ	population mean (4)
N	number of subjects or observations (1)
N_G	number of observations or subjects in group G (15)
π	hypothetical population proportion (10)
p	observed sample proportion (10)
$P(A)$	probability of event A (8)
PR	percentile rank (3)
ϕ	phi coefficient (17)
r_C	matched-pairs rank biserial correlation (18)
r_G	Glass rank biserial correlation (18)
r_{XY}	sample Pearson correlation coefficient between X and Y (12)
r_s	Spearman rank-order correlation coefficient (13)
r_{pb}	point-biserial correlation coefficient (14)
$\bar{R}_i$	mean of ranks in group i (18)
ρ_{XY}	population correlation coefficient between X and Y (10)
s	sample standard deviation (5)
s^2	population variance estimate (5)
s_D^2	variance of the Ds (11)
s^2_{pooled}	pooled variance (11)
$s_{\bar{X}}$	standard error of the mean (10)
$s_{\bar{X}_1-\bar{X}_2}$	standard error of the difference (11)
$s_{Y'}$	estimate of σ_{Y}' obtained from a sample (12)
$Score_p$	score corresponding to the pth percentile (3)
SFB	sum of frequencies below the critical interval (3)
SS	sum of squares (15)
SS_B	sum of squares between groups (15)
SS_T	total sum of squares (15)
SS_W	sum of squares within groups (15)
SS_1	sum of squares for factor 1 (16)
SS_2	sum of squares for factor 2 (16)

$SS_{1 \times 2}$	sum of squares for interaction (16)
$\sum$	sum or add up (1)
σ	standard deviation (5)
σ^2	variance (5)
σ_p	standard error of a sample proportion (10)
$\sigma_{Y'}$	standard error of estimate for predicting Y (12)
t	statistic following the t distribution (10)
T	T score (6)
T_E	expected sum of ranks (18)
T_i	sum of ranks in group i (18)
θ	theta (14)
X'	predicted X score (12)
$\bar{X}$	sample mean (4)
$\bar{X}_G$	mean of group G (15)
Y'	predicted Y score (12)
z	standard score based on a normal distribution (9)
Z	standard score (6)

Algebra Review

In order to master the elementary course in statistics for which this text is designed, you must have a working knowledge of basic mathematical and algebraic principles. This section provides a brief refresher of some of these concepts and operations, including symbols, fractions, exponents, square roots, and factoring. Although some students will not need to spend much time with this section, *no* student should leave it until thoroughly familiar with its material.

SYMBOLS

You are undoubtedly familiar with the signs for equality (=) and inequality (≠), and for addition (+), subtraction (−), multiplication (× or ·), and division (÷). Some other symbols often found in statistics are >, which means "is greater than"; and <, which means "is less than." For example, $5 > 2$ is read as "5 is greater than 2." Conversely, $2 < 5$ is read as "2 is less than 5."

Sometimes it is necessary to write that some quantity is "greater than or equal to" some other quantity. This can be expressed by the symbol ≥, which is a combination of the signs > and =. $M \geq 10$ is therefore read as "M is greater than or equal to 10." Similarly, the symbol ≤ means "is less than or equal to."

Another symbol commonly found in statistics is $|x|$, which is read "the absolute value of x." This means that whether x is a positive or a negative number, you are to take its positive (or absolute) value. For example, $|-7| = 7$; $|7| = 7$.

POSITIVE AND NEGATIVE NUMBERS

While it can safely be assumed that any student taking a college course in statistics is familiar with the rules governing simple addition, subtraction, multiplication, and division, a problem might arise when these basic arithmetical processes involve negative numbers. It helps to think of a number line, running from negative numbers on the left, through zero, to positive numbers on the right.

$-4 \quad -3 \quad -2 \quad -1 \quad 0 \quad +1 \quad +2 \quad +3 \quad +4$

Any given number has both quantity (its distance from zero) and sign (its direction from zero).

Addition and Subtraction

Adding or subtracting positive and negative numbers can be seen as moving up or down the line a certain number of steps. If you wish to add +1 and +3, you start at +1 and move three steps to the right, ending at +4. Similarly, if you wish to add −1 and +3, you start at −1, move three steps to the right, and end at +2. In subtracting numbers, you move to the left. Since it is impractical to plot a number line every time you have to add or subtract, certain rules have been formulated and should be memorized:

1. Adding a negative number is the same as subtracting that quantity:

$$7 + (-4) = 7 - 4 = 3$$
$$4 + (-7) = 4 - 7 = -3$$

2. Subtracting a negative number is the same as adding that quantity:

$$7 - (-4) = 7 + 4 = 11$$
$$4 - (-7) = 4 + 7 = 11$$

Multiplication and Division

The inclusion of negative numbers in multiplication and division problems should cause you no trouble if you remember this rule: when numbers of different signs are to be multiplied, the product is positive if there is an even number of negative terms in the multiplication, and negative if there is an odd number.

For example:

$$(-4)\ (-2) = +8$$
$$(-4)\ (+2) = -8$$
$$(-4)\ (-2)\ (+2) = +16$$
$$(-4)\ (+2)\ (+2) = -16$$
$$(-x)\ (-y) = xy$$
$$(-x)\ (+y) = -xy$$

The same rule applies to division. If there is an even number of negative terms, the result is positive; if there is an odd number of negative terms, the result is negative.

For example:

$$\frac{-4}{-2} = 2$$

$$\frac{-4}{2} = -2$$

$$\frac{(-x)\,(-y)}{z} = \frac{xy}{z}$$

$$\frac{(-x)\,(y)}{z} = \frac{xy}{z}$$

FRACTIONS

Reducing Fractions

Computations can be simplified by *reducing* the fractions involved. This is done by dividing both the numerator and the denominator by the same number, one that goes evenly into both:

$$\frac{9}{15} = \frac{9 \div 3}{15 \div 3} = \frac{3}{5}$$

Common Denominator

Fractions can be added or subtracted only if they have—or if we give them—a common denominator. For example:

$$\frac{2}{5} + \frac{1}{4}$$

The obvious common denominator for these two fractions is 20, since it is the smallest number which both denominators go into evenly. 5 goes into 20 four times, therefore both the numerator and the denominator of the first fraction are multiplied by 4:

$$\frac{2 \times 4}{5 \times 4} = \frac{8}{20}$$

Four goes into 20 five times, so both the numerator and the denominator of the second fraction are multiplied by 5:

$$\frac{1 \times 5}{4 \times 5} = \frac{5}{20}$$

Addition and Subtraction

Now the fractions can be added. The numerators are added while the denominator is maintained:

$$\frac{8}{20} + \frac{5}{20} = \frac{13}{20}$$

Had this been a subtraction problem rather than addition, the process would have been the same with the obvious exception that the second numerator would have been subtracted from rather than added to the first, i.e.:

$$\frac{2}{5} - \frac{1}{4} = \frac{8}{20} - \frac{5}{20} = \frac{3}{20}$$

Multiplication and Division

To multiply fractions, simply multiply the numerators to get the numerator product and multiply the denominators to get the denominator product:

$$\frac{3}{4} \times \frac{5}{6} = \frac{15}{24}$$

This can be reduced to $5/8$.

Dividing fractions is also simple. All you need do is invert the divisor (the bottom fraction) and then multiply by the other fraction:

$$\frac{\frac{3}{4}}{\frac{5}{6}} = \frac{3}{4} \times \frac{6}{5} = \frac{18}{20} \text{ which can be reduced to } \frac{9}{10}.$$

SOLVING FOR AN UNKNOWN

Most equations observed in statistics involve an unknown value. For example:

$$5 + x = 7$$

with x as the unknown quantity whose value must be found. We solve for x by isolating it on one side of the equation. In isolating an unknown, it is essential that the equality of the two sides of the equation be maintained.

We do this by always treating both sides of the equation identically. If a number is subtracted from or added to one side of an equation, that same number must be subtracted from or added to the other side. This principle also holds true if one side of an equation is being multiplied or divided by a number: the other side must also be multiplied or divided by that same number.

To go back to our example:

$$5 + x = 7$$

To isolate x on one side of the equation, we must subtract 5 from $5 + x$. To maintain the equality, we must also subtract 5 from the other side of the equation:

$$5 + x - 5 = 7 - 5$$
$$x = 2$$

As another example, take $3x = 9$. We isolate x by dividing each side of the equation by 3:

$$3x \div 3 = 9 \div 3$$
$$x = 3$$

As a general rule, any mathematical operation that is done to one side of an equation can be done to the other and the equality will be maintained.

EXPONENTS

An exponent is a number or symbol placed as a superscript to a second quantity known as the base number. The exponent signifies that the base number should be multiplied by itself a specified number of times:

$$3^4 = 3 \times 3 \times 3 \times 3 = 81$$

It should be noted that $x^1 = x$ and $x^0 = 1$.

Addition and Subtraction

Numbers with exponents cannot be added or subtracted without first performing the indicated exponentiation. This is true even if the base numbers are the same. For example, $2^3 + 2^4$ is determined by carrying out the exponentiation and then adding:

$$2^3 + 2^4 = (2 \times 2 \times 2) + (2 \times 2 \times 2 \times 2) = 8 + 16 = 24$$

(In contrast: $2^7 = 2 \times 2 \times 2 \times 2 \times 2 \times 2 \times 2 = 128$!)

Multiplication and Division

However, in multiplying, the exponents *can* be added if the base numbers are the same:

$$2^3 \times 2^4 = 2^{3+4} = 2^7 = 128$$

Also, in dividing, the exponents can be subtracted if the base numbers are the same:

$$\frac{3^4}{3^2} = 3^{4-2} = 3^2 = 9$$

Or:

$$\frac{x^y}{x^z} = x^{y-z}$$

If the exponents do not have the same base number (e.g.: $3^4 \times 5^2$), then the exponentiation *must* be carried out first.

Whenever a number has a negative exponent, a reciprocal (1/number) is taken. For example:

$$3^{-2} = \frac{1}{3^2} = \frac{1}{3 \times 3} = \frac{1}{9}$$

SQUARE ROOTS

There is not sufficient space here to review the process of extracting a square and, since most students have calculators, it is not really necessary. However, we will indicate how the radical (i.e.: the square root sign) is treated in various mathematical operations. In brief, the radical can be applied to each number in a complex term that requires multiplication or division, but *not* to the numbers in a term requiring addition or subtraction.

For example:

$$\sqrt{16x} = \sqrt{16} \times \sqrt{x} = 4\sqrt{x}$$

$$\sqrt{\frac{x+y}{z}} = \sqrt{x+y} \div \sqrt{z}$$

but

$$\sqrt{4+16} \neq \sqrt{4} + \sqrt{16}$$

$$\sqrt{x-z} \neq \sqrt{x} - \sqrt{z}$$

ALGEBRAIC MANIPULATION

Recall from your algebra course that:

1. $a(b + c) = ab + ac$
   example: $6(3 + 1) = 6(4) = 24$
   $= (6)(3) + (6)(1)$
   $= 18 + 6 = 24$

2. $a(b - c) = ab - ac$
   example: $7(5 - 2) = 7(3) = 21$
   $= (7)(5) - (7)(2)$
   $= 35 - 14 = 21$

3. $(a + b)(c + d) = ac + ad + bc + bd$
   example: $(7 + 4)(6 + 2) = (11)(8) = 88$
   $= (7)(6) + (7)(2) + (4)(6) + (4)(2)$
   $= 42 + 14 + 24 + 8 = 88$

4. $(a - b)(c - d) = ac - ad - bc + bd$
   example: $(4 - 2)(7 - 3) = (2)(4) = 8$
   $= (4)(7) - (4)(3) - (2)(7) + (2)(3)$
   $= 28 - 12 - 14 + 6 = 8$

BINOMIAL EXPANSION

Many statistical problems require the expansion of a binomial. A binomial is an expression which involves the addition or subtraction of two quantities, for example:

$(x + y)^2$. Expanded, $(x + y)^2 = x^2 + 2xy + y^2$.

This expansion is obtained by:

1. Taking the square of the first term in the binomial $+x^2$
2. Adding two times the product of both terms $+2xy$
3. Adding the square of the second term $+y^2$

Or, using the rules in the preceding section:

$(x + y)(x + y) = xx + xy + xy + yy = x^2 + 2xy + y^2$

The same steps would have been taken had the binomial been $(x - y)^2$.

1. Taking the square of the first term $-x^2$

2. Adding two times the product of the terms of the binomial $-2xy$
3. Adding the square of the second term $-y^2$
   Therefore, $(x - y)^2 = x^2 - 2xy + y^2$
   Or, using the rules in the preceding section:

   $$(x - y)(x - y) = xx - xy - xy + yy = x^2 - 2xy + y^2$$

Hypothetical Scores on a 20-Point Psychology Test for Students Drawn at Random from Four Universities

The following (fictitious) data will be used in various problems in this workbook.

UNIVERSITY A	17	12	6	13	9	15	11	16	4	15
($N = 50$)	12	13	10	13	2	11	13	10	20	14
	12	17	10	15	12	17	9	14	11	15
	9	18	12	13	12	17	8	16	12	15
	11	16	9	13	18	10	13	0	11	16
UNIVERSITY B	17	8	12	12	3	12	7	14	1	11
($N = 50$)	12	11	9	14	10	13	7	13	8	12
	9	12	17	11	6	10	10	3	9	8
	6	13	5	16	10	9	19	5	12	10
	16	11	14	11	13	12	2	17	10	14
UNIVERSITY C	9	11	6	5	4	9	0	4	5	7
($N = 10$)										
UNIVERSITY D	14	8	17	6	10					
($N = 5$)										

1. Summation Notation

REMINDER $\sum X$ is a shorthand version of

$$\sum_{i=1}^{N} X_i$$

($N =$ total number of subjects or cases).

1. $\sum (X+Y) = \sum X + \sum Y$
2. $\sum (X-Y) = \sum X - \sum Y$
3. $\sum XY$ (multiply first, then add) $\neq \sum X \sum Y$ (add first, then multiply)
4. $\sum X^2$ (square first, then add) $\neq (\sum X)^2$ (add first, then square)

If k is a constant,

5. $\sum k = Nk$
6. $\sum (X+k) = \sum X + Nk$
7. $\sum (X-k) = \sum X - Nk$
8. $\sum kX = k \sum X$

PROBLEMS

1. Express the following words in symbols.

a. Add up all the scores on test X, then add up all the scores on test Y, and then add the two sums together.

b. Add up all the scores on test G. To this, add the following: the sum obtained by squaring all the scores on test P and then adding them up.

c. Square all scores on test X. Add them up. From this, subtract 6 times the sum you get when you multiply each score on X by the corresponding score on Y and add them up. To this, add 4 times the quantity obtained by adding up all the scores on test X and squaring the result. To this, add twice the sum obtained by squaring each Y score and then adding them up. (Compare the amount of space needed to express this equation in words with the amount of space needed to express it in symbols. Do you see why summation notation is necessary?)

2. Five students are enrolled in an advanced course in psychology. Two quizzes are given early in the semester, each worth a total of ten points. The results are as follows:

student	*quiz 1* (X)	*quiz 2* (Y)
1	0	2
2	2	6
3	1	7
4	3	6
5	4	9

a. Compute each of the following:

$\sum X =$ ________ $(\sum X)^2 =$ ________ $\sum (X - Y) =$ ________

$\sum Y =$ ________ $(\sum Y)^2 =$ ________ $\sum X - \sum Y =$ ________

$\sum X^2 =$ ________ $\sum (X + Y) =$ ________ $\sum XY =$ ________

$\sum Y^2 =$ ________ $\sum X + \sum Y =$ ________ $\sum X \sum Y =$ ________

$\sum_{i=1}^{3} X_i =$ ________ $\sum_{i=2}^{5} Y_i =$ ________ $\sum_{i=2}^{4} X_i Y_i =$ ________

b. Using the above results, show that each of the following rules given in the reminder for this chapter is true:

Rule 1: ________ = ________.

Rule 2: ________ = ________

Rule 3: ________ $\neq$ ________

Rule 4: ________ $\neq$ ________ (X data)

________ $\neq$ ________ (Y data)

c. After some consideration, the instructor decides that Quiz 1 was excessively difficult and decides to add four points to each student's score. This can be represented in symbols by using k to stand for the constant amount in question, 4 points.

Using rule 6, compute $\sum (X + k) =$ ________ + ________ = ________.

Compute $\sum X + k =$ ________ + ________ = ________. (Note that this result is different from the preceding one.)

Now add four points to each student's score on Quiz 1 and obtain the sum of these new scores.

Sum = ________

d. Had the instructor been particularly uncharitable, he might have decided that Quiz 2 was too easy and subtracted three points from each student's score on that quiz. Since this is a new problem, the letter k can again be used to represent the constant; here, $k = 3$.

Using rule 7, compute $\sum (Y - k) =$ ________ $-$ ________ $=$ ________.

Compute $\sum Y - k =$ ________ $-$ ________ $=$ ________. (Note that this result is different from the preceding one.)

Now subtract three points from each student's score on Quiz 2 and obtain the sum of these new scores.

Sum = ________

e. Suppose that the instructor decides to double all of the original scores on Quiz 1.

Using rule 8, compute $\sum kX =$ ________ $\cdot$ ________ $=$ ________.

Now double each student's score on Quiz 1 and obtain the sum of these new scores.

Sum = ________

3. Using the data given on page 1 of this workbook, compute the following:

a. For University C:

$\sum X =$ ________ $\sum X^2 =$ ________ $(\sum X)^2 =$ ________

b. For University D:

$\sum X =$ ________ $\sum X^2 =$ ________ $(\sum X)^2 =$ ________

If you would like some additional practice, you may verify that:

For University A, $\sum X = 617$; $\sum X^2 = 8385$; $(\sum X)^2 = 380{,}689$

For University B, $\sum X = 526$; $\sum X^2 = 6316$; $(\sum X)^2 = 276{,}676$

ADDITIONAL COMPUTATIONAL PRACTICE

For each of the following (separate) sets of data, compute the values needed in order to fill in the answer spaces. Then answer the additional questions that follow.

data set 1:

S	X	Y
1	1	2
2	3	5
3	1	0
4	0	1
5	2	3

$N =$ ________

$\sum X =$ ________ $\sum Y =$ ________

$\sum X^2 =$ ________ $\sum Y^2 =$ ________

$(\sum X)^2 =$ ________ $(\sum Y)^2 =$ ________

$\sum XY =$ ________ $\sum X \sum Y =$ ________

$\sum(X + Y) =$ ________ $\sum(X - Y) =$ ________

data set 2:

S	X	Y
1	7.14	0
2	8.00	2.60
3	0	4.32
4	4.00	2.00
5	4.00	6.00
6	1.00	1.15
7	2.25	1.00
8	10.00	3.00

$N =$ ________

$\sum X =$ ________ $\sum Y =$ ________

$\sum X^2 =$ ________ $\sum Y^2 =$ ________

$(\sum X)^2 =$ ________ $(\sum Y)^2 =$ ________

$\sum XY =$ ________ $\sum X \sum Y =$ ________

$\sum(X + Y) =$ ________ $\sum(X - Y) =$ ________

data set 3:

S	X	Y
1	97	89
2	68	57
3	85	87
4	74	76
5	92	97
6	92	79
7	100	91
8	63	50
9	85	85
10	87	84
11	81	91
12	93	91
13	77	75
14	82	77

$N =$ ______

$\sum X =$ ______ $\sum Y =$ ______

$\sum X^2 =$ ______ $\sum Y^2 =$ ______

$(\sum X)^2 =$ ______ $(\sum Y)^2 =$ ______

$\sum XY =$ ______ $\sum X \sum Y =$ ______

$\sum(X+Y) =$ ______ $\sum(X-Y) =$ ______

	set 1	*set 2*	*set 3*
If every X score is multiplied by 3.2, what is the new $\sum X$ in each set?	______	______	______
If 7 is subtracted from every Y score, what is the new $\sum Y$ in each set?	______	______	______
If 1.8 is added to every X score, what is the new $\sum X$ in each set?	______	______	______
If every Y score is divided by 4, what is the new $\sum Y$ in each set?	______	______	______

(Hint: Use the appropriate summation rule in each case so as to make the calculations easier.)

2. Frequency Distributions and Graphs

REMINDER 1. Regular frequency distributions

List every score value in the first column, with the highest score at the top. List the *frequency* (symbolized by f) of each score to the right of the score in the second column.

2. Grouped frequency distributions

List the *class intervals* in the first column and the frequencies in the second column. It is usually desirable to:

1. Have a total of from 8 to 15 class intervals.
2. Use an interval size of 2, 3, 5, or a multiple of 5, selecting the smallest size that will satisfy the first rule. (All intervals should be the same size.)
3. Make the lowest score in each interval a multiple of the interval size.

Do not use grouped frequency distributions if all scores can be quickly and conveniently reported, because grouped frequency distributions lose information.

3. Cumulative frequency distributions

To the right of the frequency column, form a column of *cumulative frequencies* (symbolized by cf) by starting with the frequency for the lowest score and adding up the frequencies as you go along.

4. Graphic representations

Histograms, in which the frequency of any score is expressed by the height of the bar above that score, are particularly appropriate for discrete data (where results between the score values shown cannot occur).

Frequency polygons, in which the frequency of any score is expressed by the height of the point above that score and points are connected by straight lines, are particularly appropriate for continuous data (where results between the score values shown can occur, or could if it were possible to measure with sufficient refinement).

Stem-and-leaf displays combine features of the frequency distribution and the histogram. Class intervals ("stems") are listed in a column at the left, with the specific values within each interval listed on a horizontal line next to that interval (often represented solely by the units digit).

PROBLEMS

1. Make up a separate regular frequency distribution for each of the four universities. After you have made up the regular frequency distributions, make up a cumulative frequency distribution for each university.

University A			*University B*			*University C*			*University D*		
Score	*f*	*cf*	*Score*	*f*	*cf*	*Score*	*f*	*cf*	*Score*	*f*	*cf*

2. Make up a separate grouped frequency distribution for University A and University B. After you have made up the grouped frequency distributions, make up a cumulative frequency distribution for each university.

University A			*University B*		
class interval	*f*	*cf*	*class interval*	*f*	*cf*

3. Smedley Trueblood, a struggling young student who is taking introductory statistics for the fourth time, decides in a burst of enthusiasm to make up grouped frequency distributions for University C and University D. Is this a good idea or a bad idea? Why?

4. Plot a histogram of the *grouped* data for University A on one sheet of graph paper.

5. Plot frequency polygons for the *grouped* frequency distributions for both University A and University B on a single graph. In accordance with standard practice, use the horizontal axis (X axis) to represent test scores and the vertical axis (Y axis) to represent f.

6. Plot the cumulative frequency distributions based on the *grouped* data for both University A and University B on a single graph, using the horizontal axis to represent test scores and the vertical axis to represent *cf*.

7. Make up a stem-and-leaf display for University A, using an interval size of 3.

8. For each of the frequency distributions shown below, state whether it is:

a. (approximately) normal
b. unimodal, skewed to the right
c. unimodal, skewed to the left
d. bimodal, approximately symmetric
e. bimodal, skewed to the right
f. bimodal, skewed to the left
g. (approximately) rectangular
h. J-curve

Write the letter corresponding to the correct answer in the space provided beneath each frequency distribution.

(1)		(2)		(3)		(4)		(5)		(6)	
X	f	X	f	X	f	X	f	X	f	X	f
10	0	45–49	2	27–29	2	55–59	0	10	3	80–89	5
9	0	40–44	3	24–26	8	50–54	1	9	4	70–79	4
8	1	35–39	1	21–23	17	45–49	3	8	10	60–69	4
7	0	30–34	4	18–20	24	40–44	3	7	6	50–59	3
6	1	25–29	2	15–17	16	35–39	8	6	2	40–49	5
5	3	20–24	5	12–14	6	30–34	13	5	5	30–39	6
4	0	15–19	12	9–11	8	25–29	19	4	11	20–29	5
3	2	10–14	9	6–8	2	20–24	12	3	6	10–19	4
2	6	5–9	3	3–5	3	15–19	10	2	2	0–9	4
1	14	0–4	0	0–2	1	10–14	4	1	1		
0	21					5–9	2	0	0		
						0–4	0				
Ans: ______		Ans: ______		Ans: ______		Ans: ______		Ans: ______		Ans: ______	

3. Transformed Scores I: Percentiles

REMINDER

1. CASE 1. **Given a score, find the corresponding percentile rank** (*PR*)

1. Find the class interval in which the score falls.

2. Set up the following diagram and fill in the missing values:

	f	Percent ($= f/N$)
All higher intervals	....	$H\% =$ ....
Interval *in* which score falls	....	$I\% =$ ....
All lower intervals	....	$L\% =$ ....
		Check = 100%

3. Let

LRL = *lower real limit* of the interval in which the score falls. (The lower real limit of an interval is the number halfway between the lowest number in that interval and the highest number in the next lower interval.)

h = interval size

Then,

$$PR = L\% + \left(\frac{Score - LRL}{h} \cdot I\%\right)$$

2. CASE 2. **Given a percentile** (p), **find the corresponding raw score** ($Score_p$)

1. Compute $p \times N = N$th case. Find the interval in which this case falls.

2. Let LRL = lower real limit of this interval
SFB = sum of frequencies below this interval
f = frequency within this interval
h = interval size

Then,

$$Score_p = LRL + \frac{pN - SFB}{f} h$$

illustrative examples

CASE 1. *Example 1.* A student at University A scored 15 on the 20-point psychology test. Using the *grouped* data (Chapter 2, problem 2), find his percentile rank.

1. The score falls in the 14–15 interval.

2. Diagram:

	f	Percent
All scores above 14–15 interval	11	$H\% = 22\%$
14–15 interval	7	$I\% = 14\%$
All scores below 14–15 interval	32	$L\% = 64\%$
		Check = 100%

3. $LRL = 13.5 \qquad h = 2$

$$PR = 64\% + \left(\frac{15 - 13.5}{2} \cdot 14\%\right) = 64\% + 10.5\% = \underline{74.5\%}$$

Example 2. Compute the percentile rank for the student with a score of 15 from the *ungrouped* data (Chapter 2, problem 1).

1. The score falls in the 15 "interval."

2. Diagram:

	f	Percent
All scores above 15	11	$H\% = 22\%$
Scores of 15	5	$I\% = 10\%$
All scores below 15	34	$L\% = 68\%$
		Check = 100%

3. $LRL = 14.5 \qquad h = 1$

$$PR = 68\% + \left(\frac{15 - 14.5}{1} \cdot 10\%\right) = 68\% + 5\% = \underline{73\%}$$

In words, the student did fairly well and is just outside the top 25% of his class.

CASE 2. *Example 1.* A student at University A has a percentile rank of 31. Using the *grouped* data (Chapter 2, problem 2), what is his raw score?

1. $p = .31 \qquad p \times N = .31 \times 50 = 15.5$th case, which falls in the 10–11 interval.

2. $LRL = 9.5 \qquad SFB = 9 \qquad f = 9 \qquad h = 2$

$$Score_{.31} = 9.5 + \frac{15.5 - 9}{9}\,2 = 9.5 + 1.4 = \underline{10.9}$$

In words, a student who places at the 31st percentile has a raw score of approximately 11.

Example 2. Using the ungrouped data (Chapter 2, problem 1), what is the minimum score a student must have to be in the top 10% of the group from University A?

1. "Top 10%" means 90th percentile; thus $p = .90$
$p \times N = .90 \times 50 = 45$th case, which falls in the 17 "interval."

2. $LRL = 16.5 \qquad SFB = 43 \qquad f = 4 \qquad h = 1$

$$Score_{.90} = 16.5 + \frac{45 - 43}{4}\,1 = 16.5 + .5 = \underline{17}$$

In words, a student must have a score of 17 or more to be in the top 10% of the group from University A.

PROBLEMS

1. For the *ungrouped* data for University B (Chapter 2, problem 1), what is the percentile rank corresponding to a score of:

a. 5?

$L\% =$ ____________

$LRL =$ ____________

$h =$ ____________

$I\% =$ ____________

$PR =$ ____________

b. 10?

$L\% =$ ____________

$LRL =$ ____________

$h =$ ____________

$I\% =$ ____________

$PR =$ ____________

2. Again for the ungrouped data for University B, what score corresponds to:

a. the 25th percentile?

$LRL =$ ____________

$SFB =$ ____________

$f =$ ____________

$h =$ ____________

$Score_{.25} =$ ____________

b. the 75th percentile?

$LRL =$ ____________

$SFB =$ ____________

$f =$ ____________

$h =$ ____________

$Score_{.75} =$ ____________

3. Using the *grouped* data for University B, what is the percentile rank corresponding to a score of:

a. 16?

$L\% =$ ________

$LRL =$ ________

$h =$ ________

$I\% =$ ________

$PR =$ ________

b. 7?

$L\% =$ ________

$LRL =$ ________

$h =$ ________

$I\% =$ ________

$PR =$ ________

4. Again using the grouped data for University B, what score corresponds to:

a. the third decile?

$LRL =$ ____________

$SFB =$ ____________

$f =$ ____________

$h =$ ____________

$Score_{.30} =$ ____________

b. the 60th percentile?

$LRL =$ ____________

$SFB =$ ____________

$f =$ ____________

$h =$ ____________

$Score_{.60} =$ ____________

5. Smedley Trueblood's sister Susanne, who is equally confused by statistics, is delighted because she has scored at the 93rd percentile on test *X* while Smedley has scored at the 80th percentile on test *Y*. She concludes that she has outperformed Smedley. Is her conclusion justified?

ADDITIONAL COMPUTATIONAL PRACTICE

For the set of data shown below, fill in the missing values in the first table by changing every score from 85 to 99 (inclusive) to a percentile rank. Then complete the second table by computing the score corresponding to each percentile included in the table.

X	f	cf
100	0	37
99	1	37
98	0	36
97	1	36
96	0	35
95	2	35
94	3	33
93	4	30
92	8	26
91	6	18
90	5	12
89	2	7
88	3	5
87	1	2
86	0	1
85	1	1
below 85	0	0

Table 1 *Percentile ranks*

score	*H%*	*I%*	*L%*	*PR*
99	________	________	________	________
98	________	________	________	________
97	________	________	________	________
96	________	________	________	________
95	________	________	________	________
94	________	________	________	________
93	________	________	________	________
92	________	________	________	________
91	________	________	________	________
90	________	________	________	________
89	________	________	________	________
88	________	________	________	________
87	________	________	________	________
86	________	________	________	________
85	________	________	________	________

Table 2 *Score$_p$ values*

p	pN	LRL	SFB	f	$score_p$
.99					
.90					
.80					
.75					
.70					
.60					
.50					
.40					
.30					
.25					
.20					
.10					
.01					

4. Measures of Central Tendency

REMINDER 1. The sample mean

$$\bar{X} = \frac{\sum X}{N}$$

1. Use the formula $\bar{X} = (\sum fX)/N$ with regular frequency distributions.

2. The sample mean ($\bar{X}$) is an estimate of the population mean (μ).

2. The median

For either grouped or ungrouped data, compute the score corresponding to the 50th percentile. Recall from Chapter 3 that:

$$Score_{.50} = Mdn = LRL + \frac{.50N - SFB}{f} h$$

3. The mode

The most frequently obtained score. For grouped frequency distributions, the mode is the midpoint of the class interval for which f is the largest.

Use the median for highly skewed data or for extreme observations whose exact values are not known. Otherwise, use the mean. The mode is at best a rough measure and is generally less useful than the mean or median.

PROBLEMS

1. **a.** Using the results of Chapter 1, problem 3, compute the *mean* of:

University A: $\bar{X}_A =$ ....

University B: $\bar{X}_B =$ ....

University C: $\bar{X}_C =$ ....

University D: $\bar{X}_D =$ ....

b. At each university, several hundred students took the psychology quiz. Would it be correct to draw conclusions about which of the *populations* (that is, *all* students at a given university who took the psychology quiz) were better or worse, based solely on these *sample* means? Why or why not?

2. Using the formula $\bar{X} = (\sum fX)/N$, compute the mean for University C from the *ungrouped* data (Chapter 2, problem 1). Compare this answer with the mean for University C computed in problem 1 of this chapter.

$\sum fX =$ ____________

$\bar{X} =$ ____________

3. Compute the median for University A, using the *ungrouped* data (Chapter 2, problem 1).

$LRL =$ ....

$SFB =$ ....

$f =$ ....

$h =$ ....

$Mdn =$ ....

4. The unfortunate Smedley Trueblood strongly objects to the procedure in problem 3 of this chapter. He argues that the median should always be computed from *grouped* frequency distributions because grouping saves a great deal of time and trouble. What is your opinion about this?

5. Using the *ungrouped* data (Chapter 2, problem 1), what is the mode of University B?

Mode = ...

6. Using the *ungrouped* data for University C, show numerically that $\sum (X - \bar{X}) = 0$.

7. In each of the following problems, select and compute the appropriate measure of central tendency.

a. A clinical psychologist wishes a measure of central tendency for the number of trials taken to learn a task by mentally defective individuals. If a person has not learned the task after 50 trials, the psychologist ends the experiment for that individual.

Person 1 : 25 trials
Person 2 : 22 trials
Person 3 : 25 trials
Person 4 : 44 trials
Person 5 : 24 trials
Person 6 : 25 trials
Person 7 : 38 trials
Person 8 : did not learn after 50 trials
Person 9 : did not learn after 50 trials

b. A psychologist is intrigued by the fact that departmental meetings always seem to start late, and decides to collect data to find out just how late the average meeting starts. In one academic year, he attends 15 meetings, and the results are as follows:

minutes late	*f*
more than 10	0
10	1
9	0
8	1
7	1
6	0
5	3
4	3
3	2
2	2
1	1
0	1
early	0

c. In order to find out how bad things actually are, Smedley Trueblood decides to compute his grade-point average, where A = 4 points, B = 3 points, C = 2 points, D = 1 point, and F = 0 points. All courses that he has taken are worth the same amount of credit, and each F in statistics counts as a full course grade. His grades are:

statistics (first try) : F
statistics (second try) : F
statistics (third try) : F
English 1 : C
English 2 : C
introductory psychology : D
Biology 1 : C
Biology 2 : C
calculus 1 : D
calculus 2 : D
basket-weaving : A
economics 1 : B
economics 2 : C
French 1 : C
French 2 : C

8. a. In a *positively skewed* distribution, which is larger, the mean or the median?

b. In a *negatively skewed* distribution, which is larger, the mean or the median?

c. In a *symmetric* distribution (such as the normal distribution), which is larger, the mean or the median?

d. If, in the data for problem 7b, the number 10 is changed to 225,000,000,000, but the other numbers remain the same, what is the effect on the *mean* and the *median*?

e. If, in the data for problem 7b, the number 10 is changed to 0, but the other numbers remain the same, what is the effect on the *mean* and the *median*?

9. Susanne Trueblood is asked to compute the mean of the following set of data:

X	f
5	2
4	3
3	6
2	2
1	0
0	1

Her answer is:

$$\bar{X} = \frac{5+4+3+2+1+0}{6}$$

$$= \frac{15}{6}$$

$$= 2.5$$

Is this correct? If not, what error did she make?

ADDITIONAL COMPUTATIONAL PRACTICE

For each of the following (separate) sets of data, compute the values needed to fill in the answer spaces. Then compute the mean of X and the mean of Y for each of the three sets of data in the "Additional Computational Practice" problems for Chapter 1.

data set 1	*data set 2*	*data set 3*	
X	W	X	f
1	68.36	10	0
3	15.31	9	1
6	77.42	8	2
0	84.00	7	4
1	76.59	6	6
1	68.43	5	7
2	72.41	4	5
1	83.05	3	2
4	91.07	2	1
	80.62	1	0
	77.83	0	1

$\bar{X} =$ ________ $\quad \overline{W} =$ ________ $\quad \bar{X} =$ ________

$Mdn =$ ________

Chapter 1 "Additional Computational Practice" problems

	$\bar{X}$	$\bar{Y}$
data set 1	____________	____________
data set 2	____________	____________
data set 3	____________	____________

5. Measures of Variability

REMINDER 1. The variance and the standard deviation

1. *Variance*

Definition:

$$\sigma^2 = \frac{\sum (X - \bar{X})^2}{N}$$

Computing formula:

$$\sigma^2 = \frac{1}{N}\left[\sum X^2 - \frac{(\sum X)^2}{N}\right]$$

Formulas for use with regular frequency distributions:

$$\sigma^2 = \frac{\sum f(X - \bar{X})^2}{N}$$

$$\sigma^2 = \frac{1}{N}\left[\sum fX^2 - \frac{(\sum fX)^2}{N}\right]$$

2. *Standard Deviation*

$$\sigma = +\sqrt{\sigma^2}$$

σ^2 and σ are basic measures of the variability of any set of data.

2. The population variance estimate

When *the data of a sample are to be used to estimate the variance of the population from which the sample was drawn,* compute the *population variance estimate.*

Definition:

$$s^2 = \frac{\sum (X - \bar{X})^2}{N - 1}$$

Computing formula:

$$s^2 = \frac{1}{N-1}\left[\sum X^2 - \frac{(\sum X)^2}{N}\right]$$

Formulas for use with regular frequency distributions:

$$s^2 = \frac{\sum f(X-\bar{X})^2}{N-1}$$

$$s^2 = \frac{1}{N-1}\left[\sum fX^2 - \frac{(\sum fX)^2}{N}\right]$$

Note that in this situation, the standard deviation $s = +\sqrt{s^2}$.

3. The range

Another measure of variability sometimes encountered in the behavioral sciences is the *range*, which is equal to the highest score minus the lowest score. However, the range is at best a rough measure and is generally less useful than the variance or standard deviation.

PROBLEMS

1. Susanne Trueblood, who has great difficulty squaring numbers and arriving at the correct answer, complains that the formula for the variance is needlessly complicated and that she would rather use the formula $\sum (X-\bar{X})/N$. What is wrong with this idea? (Hint: See Chapter 4, problem 6.)

2. In a further desperate attempt to make her life easier, Susanne argues that it is totally unnecessary to divide by N when computing the variance and that this step can be omitted. What is wrong with this idea?

3. Using the results obtained in Chapter 1, problem 3, compute s^2, s, σ^2, and σ for each of the four universities. Use the *computing* formulas.

a. University A:

$s^2 =$ ____________

$s =$ ____________

$\sigma^2 =$ ____________

$\sigma =$ ____________

b. University B:

$s^2 =$ ____________

$s =$ ____________

$\sigma^2 =$ ____________

$\sigma =$ ____________

c. University C:

$s^2 =$ ______

$s =$ ______

$\sigma^2 =$ ______

$\sigma =$ ______

d. University D:

$s^2 =$ ______

$s =$ ______

$\sigma^2 =$ ______

$\sigma =$ ______

e. Now recompute s^2 and σ^2 for University D using the *definition* formulas. Do your results agree with those obtained in part d?

$s^2 =$ ______

$\sigma^2 =$ ______

4. What is the *range* of scores for:

a. University A?

b. University B?

c. University C?

d. University D?

Why are these measures of variability *not* as informative as variances and standard deviations?

5. a. Suppose that the mean of a statistics examination is 77 and you score 85. Would you (for purposes of getting a high letter grade) prefer that the standard deviation be large or small? Why?

b. Suppose that the mean is still 77, but you score only 68. Now would you prefer that the standard deviation be large or small? Why?

c. Make up one example of your own (preferably *not* involving test scores) where an individual would prefer a *small* standard deviation.

d. Now make up one example (preferably *not* involving test scores) where an individual would prefer a *large* standard deviation.

e. The standard deviation of a set of scores is zero. What does this imply about the scores?

f. On a homework assignment, Smedley Trueblood finds that the standard deviation of a set of scores is −29. What does this imply?

ADDITIONAL COMPUTATIONAL PRACTICE

For each of the following (separate) sets of data from Chapter 4, compute σ and s. Then compute σ and s for X, and σ and s for Y, for each of the three sets of data in the "Additional Computational Practice" problems for Chapter 1. Use the computing formula for all problems.

data set 1	*data set 2*	*data set 3*	
X	W	X	f
1	68.36	10	0
3	15.31	9	1
6	77.42	8	2
0	84.00	7	4
1	76.59	6	6
1	68.43	5	7
2	72.41	4	5
1	83.05	3	2
4	91.07	2	1
	80.62	1	0
	77.83	0	1

$\sigma =$ ____________ $\sigma =$ ____________ $\sigma =$ ____________

$s =$ ____________ $s =$ ____________ $s =$ ____________

Chapter 1 "Additional Computational Practice" problems

	X	Y
data set 1	$\sigma =$ ________	$\sigma =$ ________
	$s =$ ________	$s =$ ________
data set 2	$\sigma =$ ________	$\sigma =$ ________
	$s =$ ________	$s =$ ________
data set 3	$\sigma =$ ________	$\sigma =$ ________
	$s =$ ________	$s =$ ________

6. Transformed Scores II: Z and T Scores

REMINDER 1. General transformations

operation	*effect on* $\bar{X}$	*Effect on* σ	*Effect on* σ^2
1. *Add* a constant, k, to every score	new mean $=$ old mean $+ k$	no change	no change
2. *Subtract* a constant, k, from every score	new mean $=$ old mean $- k$	no change	no change
3. *Multiply* every score by a constant, k	new mean $=$ old mean $\times k$	new $\sigma = $ old $\sigma \times k$	new $\sigma^2 = $ old $\sigma^2 \times k^2$
4. *Divide* every score by a constant, k	new mean $=$ old mean$/k$	new $\sigma = $ old σ/k	new $\sigma^2 = $ old σ^2/k^2

2. Z scores (standard scores)

$$Z = \frac{X - \bar{X}}{\sigma} \qquad \text{new mean} = 0 \qquad \text{new } \sigma = 1$$

3. T scores

$$T = 10Z + 50 \qquad \text{new mean} = 50 \qquad \text{new } \sigma = 10$$

PROBLEMS

1. The mean of a set of scores is 8 and the standard deviation is 4. What will the new mean, standard deviation, and variance be if I:

	new $\bar{X}$	new σ	new σ^2
a. add 6.8 to every number?	________	________	________
b. subtract 4 from every number?	________	________	________
c. multiply every number by 3.2?	________	________	________
d. divide every number by 4?	________	________	________
e. add 6 to every number, and then divide each new number by 2?	________	________	________

2. What are the advantages of having a set of numbers with a mean of zero and a standard deviation of one (*Z* scores)?

3. Suppose you had to compute the standard deviation of the following numbers:

6,894,227
6,894,233
6,894,229
6,894,230
6,894,237

In what way could you manipulate the data so as to make the calculations easier? (Do not compute anything; just state what you would do.)

4. For each of the following, compute the *Z* score; then compute the *T* score.

	Z score	*T* score
a. A University A student with a score of 17.	________	________
b. A University A student with a score of 11.	________	________
c. A University B student with a score of 11.	________	________
d. A University C student with a score of 9.	________	________
e. A University C student with a score of 6.	________	________
f. A University D student with a score of 6.	________	________

5. Which of each pair is better, or are they the same? (You should be able to answer by inspection.)

a. A T score of 47 and a Z score of $+0.33$ ....

b. A T score of 64 and a Z score of $+0.88$ ....

c. A T score of 42 and a Z score of -1.09 ....

d. A T score of 60 and a Z score of $+1.00$ ....

e. A T score of 50 and a raw score of 26 (mean of raw scores $= 26$, $\sigma = 9$) ....

f. A Z score of $+0.04$ and a raw score of 1092 (mean of raw scores $= 1113$, $\sigma = 137$) ....

g. A Z score of zero and a T score of 50 ....

6. a. Smedley Trueblood computes his Z score on a statistics examination and arrives at an answer of $+22.8$. Ignoring the fact that Smedley is unlikely to be above average on anything concerning statistics, what would you conclude from this result?

b. Smedley, who is not convinced as to the merits of transformed scores, argues as follows: " In problem 4 of this chapter, a University A student with a raw score of 11 winds up with different Z and T scores than a University B student with a raw score of 11. Since the raw scores were the same for both students, they should wind up with the same Z and T scores, so transformations give very misleading results." Do you agree or disagree? Why?

c. Consider the following data:

	psychology test	*English test*
Smedley's score	73	67
$\bar{X}$	81.0	77.0
σ	5.0	10.0

Smedley believes that he has done better on the psychology test for two reasons: his score is higher, and he is only eight points below average (as opposed to 10 points below average on the English test). Convert each of his test scores to a Z score. Is he right?

ADDITIONAL COMPUTATIONAL PRACTICE

For data set 3 in the "Additional Computational Practice" problems for Chapter 1, transform each subject's score on X to a Z score. Then transform each subject's score on Y to a Z score. (For convenience, first write down the previously computed means and values of σ.)

data set 3, Chapter 1

$\bar{X} =$ ______ $\bar{Y} =$ ______
$\sigma_X =$ ______ $\sigma_Y =$ ______

S	X	Y	Z_X	Z_Y
1	97	89	______	______
2	68	57	______	______
3	85	87	______	______
4	74	76	______	______
5	92	97	______	______
6	92	79	______	______
7	100	91	______	______
8	63	50	______	______
9	85	85	______	______
10	87	84	______	______
11	81	91	______	______
12	93	91	______	______
13	77	75	______	______
14	82	77	______	______

Transform each raw score in the data set below into a *Z* score, a *T* score, and a SAT score.

data set 2, chapter 4

$\overline{W}$ = ____________

σ = ____________

W	*Z*	*T*	SAT
68.36	________	________	________
15.31	________	________	________
77.42	________	________	________
84.00	________	________	________
76.59	________	________	________
68.43	________	________	________
72.41	________	________	________
83.05	________	________	________
91.07	________	________	________
80.62	________	________	________
77.83	________	________	________

7. Additional Techniques for Describing Batches of Data

REMINDER

1. The 5-number summary

This numerical summary provides a good indication of the shape of a distribution by specifying five important values:

Median	
First Quartile	Third Quartile
Lowest Score	Highest Score

2. Box-and-whisker plots

This graphic summary is particularly valuable when two or more distributions are to be compared. Score values are shown on the vertical axis, with the name of the distribution(s) on the horizontal axis. The middle half of the observations in a given distribution is presented as a vertical rectangle whose top and bottom fall at the third and first quartiles, respectively. The box is divided by a line which represents the median, and lines (whiskers) are extended up to the largest and down to the smallest observation.

3. Mean-on-spoke representations

This graphic summary is preferable to box-and-whisker plots when a distribution is approximately bell-shaped. Here again, score values are shown on the vertical axis; but the mean of a given distribution is plotted as a point, with a vertical spoke running through it that is one standard deviation long on either side.

PROBLEMS

1. Make up a 5-number summary for Universities A and B (separately), using the *ungrouped* data (Chapter 2, problem 1).

University A:

University B

2. Using the results from problem 1, make up a box-and-whisker plot that compares the distribution of scores for Universities A and B.

5. Smedley Trueblood is struggling grimly to understand box-and-whisker plots, but is also in need of assistance. What can be concluded about the distribution in question if:

a. The horizontal line representing the median is near the top of the vertical rectangle?

b. The horizontal line representing the median is near the bottom of the vertical rectangle?

c. The horizontal line representing the median is at about the middle of the vertical rectangle, and the whiskers at the top and bottom of the vertical rectangle are approximately equal in length?

Review Section I. Review of Descriptive Statistics

For each of the following problems, one of the choices in the Answer Column is the correct procedure to use; select the right procedure and compute the answer.

Answer Column

a. Regular frequency distribution
b. Grouped frequency distribution
c. Mean
d. Median
e. Mode
f. Range
g. Standard deviation
h. Z score
i. Percentile, Case 1 (PR)
j. Percentile, Case 2 ($Score_p$)

PROBLEMS

1. A college dean wishes a measure of central tendency for the numerical grades of students at his University who are on academic probation. The grades of these students are shown in the following. Although the maximum possible is 100, none of these scores are above 70 because a score of 70 or more is sufficient to keep off probation.

66 47 66 60 69 68 67 10 65 65

40 67 65 51 67 54 25 62 65 65

2. A student who obtains a test score of 81 (mean = 59.0, standard deviation = 20.0) wants to know how good her score is.

3. A statistics instructor is giving the same final examination this year as he did one year ago. He wishes to determine the minimum score needed to pass. Anyone whose score would place him in the top 80% of last year's class will pass. The examination scores from the previous year are shown below; compute the minimum score needed to pass this year's examination.

62 68 86 64 63 94 71 62 73 72 68 67 62

4. A specialist in consumer research wants to know the *variability* of the set of data shown below (number of television sets owned by 100 American families). Compute the appropriate measure.

X (*number of TV sets*)	f
11 or more	0
10	1
9	0
8	1
7	0
6	4
5	2
4	5
3	10
2	22
1	35
0	20

5. An instructor obtains the following data for Part I of this review chapter; scores represent the number correct out of the total of 10 problems. Compute the appropriate measure of central tendency.

9 6 5 4 7 8 7 5 7 8

6. The same instructor as in problem 5 wishes to estimate the population variance from the sample data. Compute the appropriate measure.

7. A student in a large class wants to express his test score as a single number that will show how good the test score is. The class data are given below; compute the appropriate measure.

student's score = 31

X	f
50	0
45–49	1
40–44	3
35–39	1
30–34	7
25–29	14
20–24	6
15–19	7
10–14	12
5–9	6
0–4	3

8. Scores of 50 students on an examination range from 20 to 90, and the teacher wants to summarize all the data in a convenient format. What procedure should he use?

Answer: ....

9. A teacher obtains scores of 75 students on a 10-item quiz and wishes to summarize all the data in a convenient format. What procedure should he use?

Answer: ....

10. A comparison of the following two sets of data (admittedly exaggerated) shows why *two* of the procedures in the answer column are inferior to other procedures. Which two, and why?

sample A		*sample B*	
X	*f*	*X*	*f*
5	1	5	4
4	0	4	3
3	0	3	3
2	0	2	2
1	0	1	2
0	19	0	6

8. Probability and the General Strategy of Inferential Statistics

REMINDER 1. Probability

Definition:

$$P = \frac{\text{number of ways the specified event can occur}}{\text{total number of possible events}}$$

Odds against an event = number of unfavorable events *to* number of favorable events.

Additive law:

$$P(A \text{ or } B) = P(A) + P(B) - P(A \text{ and } B)$$

Multiplicative law: If A and B are independent events, $P(A \text{ and then } B) = P(A) \times P(B)$

2. The general strategy of inferential statistics

Problem: A coin is flipped six times and lands heads up six times. Is the coin "fair" or "loaded"? The only way to be certain is to flip the coin an infinite number of times (measure the whole population), which is impossible. Therefore, it is necessary to *draw an inference about the population based on results obtained from a sample*.

Initial Assumption: It would be helpful to know the probability of obtaining six heads in six flips. This is equal to $[P(\text{H})]^6$, where $P(\text{H})$ is the probability of a head on one flip of the coin.

If we assume at the outset that the coin is loaded, we are in trouble. There are many possible values of $P(\text{H})$ (for example, .55, .75, 1.00) and there is no way to tell what value to use when calculating $[P(\text{H})]^6$. If, on the other hand, we assume at the outset that the coin is fair, then we are assuming that $P(\text{H}) = .50$. Now we can compute $[P(\text{H})]^6$.

Procedure: 1. Assume coin is fair and $P(\text{H}) = \frac{1}{2}$.
2. Compute P (6 heads) $= [P(\text{H})]^6 = (\frac{1}{2})^6 = .016$.
3. *If* the coin is fair, the probability of obtaining six heads in six flips is .016, or about 3 in 200. According to conventional rules this is unlikely enough for us to *reject* the initial assumption that the coin is fair.

Conclusion: Therefore, our best guess is that the coin is "loaded."

PROBLEMS

1. Compute each of the following probabilities. Questions refer to the throw of one fair die. Answers may be left in fractional form.

a. The probability of throwing either a 2, 4, 5, or 6 on one throw.

b. The probability of throwing either a 2 or 3, and then a 5.

c. The probability of throwing a 3 and then a 14.

d. The odds against throwing a 6.

2. In the following problems, cards are drawn from a standard 52-card deck. Before the next draw, the card drawn is replaced and the deck is thoroughly shuffled. Compute each of the following probabilities; answers may be left in fractional form.

a. The probability of drawing a ten on the first draw.

b. The probability of drawing either a deuce, three, four, five, heart, or a diamond on the first draw.

c. The probability of drawing the ace of spades twice in a row.

d. The probability of drawing either a jack, ten, seven of clubs, or a spade on the first draw, *and then* drawing either the ace of diamonds or nine of hearts on the second draw.

3. Now suppose that instead of replacing the cards in the deck and shuffling thoroughly, any card drawn on the first draw is thrown away. Under this new procedure, recompute:

a. Problem 2c:

b. Problem 2d:

4. Smedley Trueblood visits Las Vegas and decides to play the roulette wheel, which consists of 38 different numbers (1 through 36 inclusive, 0, and 00). If he bets on a single number and that number comes up, he collects 35 times his bet. Should he expect to win, break even, or lose in the long run? Why?

5. The following questions refer to the throw of one die.

a. What is the probability of obtaining an odd number on one throw *if* the die is fair?

b. What is the probability of obtaining seven odd numbers in seven throws *if* the die is fair?

c. Could you determine the answers to parts a and b if the die is *not* fair? If not, why not?

d. Suppose that you wish to conduct an "experiment" to decide whether or not to bet with the owner of this die. Would your initial hypothesis be that the die is fair or that it is loaded? Why?

e. Suppose that the owner of the die wins if an odd number comes up and you win if an even number appears. If you obtained seven odd numbers in seven throws, what would your decision about the fairness of the die be?

f. How might your decision in part e be wrong?

6. Suppose that we wish to investigate whether or not a coin is fair, but that our "experiment" consists of flipping the coin only three times.

a. Calculate the statistical model for this three-flip experiment, and summarize the results in the table below. (Hint: there are three possible ways to obtain 2 heads and 1 tail, namely THH, HTH, and HHT. There are also three possible ways to obtain 1 head and 2 tails, namely HTT, THT, and TTH.)

Event	*P*
3 heads	______
2 heads	______
1 head	______
0 heads	______

b. What assumption must you make in order to carry out the calculations in part a?

c. What should be the sum of the probabilities you computed in part a?

d. If you plan to use the .05 decision rule, why is this experiment badly designed?

e. What design error in behavioral science research is similar to the error discussed in part d?

ADDITIONAL COMPUTATIONAL PRACTICE

A. One hundred slips of paper bearing the numbers from 1 to 100, inclusive, are placed in a large hat and thoroughly mixed. What is the probability of drawing:

1. The number 17? ________
2. The number 92? ________
3. Either a 2 or a 4? ________
4. A number from 7 to 11, inclusive? ________
5. A number in the 20's? ________
6. An even number? ________
7. An even number, and then a number from 3 to 19, inclusive? (The number drawn first is replaced prior to the second draw.) ________
8. A number from 96 to 100, inclusive, or a number from 70 to 97, inclusive? ________

B. Slips of paper are placed in a large hat and thoroughly mixed. Ten slips bear the number 1, 20 slips bear the number 2, 30 slips bear the number 3, and 5 slips bear the number 4. What is the probability of drawing:

1. A 1? ________
2. A 2? ________
3. A 3? ________
4. A 4? ________
5. A 1 or a 4? ________
6. A 1 or a 2 or a 3 or a 4? ________
7. A 5? ________
8. A 2 and then a 3? (The number drawn first is *not* replaced prior to the second draw. Answer may be left in fractional form.) ________

C. What is the probability of getting two heads in two flips of a coin if:

1. The coin is fair? ________
2. The coin is loaded, and $P(H) = .6$? ________
3. The coin is loaded, and $P(H) = .9$? ________
4. The coin is loaded, and $P(H) = 1.0$? ________
5. The coin is loaded, and $P(T) = .9$? ________
6. The coin is loaded, and $P(T) = 1.0$? ________

D. An imperfect 52-card deck contains five aces of spades and no king of spades, no queen of spades, no jack of spades, and no ten of spades, but is otherwise normal. What is the probability of drawing (answers may be left in fractional form):

1. The ace of spades? ________
2. The king of spades? ________
3. The six of hearts? ________
4. The ace of spades twice in a row? (The card drawn first is replaced prior to the second draw.) ________
5. The ace of spades and then a king? (The card drawn first is replaced prior to the second draw.) ________
6. The ace of spades or a six? ________
7. An ace or a heart? ________

9. The Normal Curve Model

REMINDER

1. TYPE 1 PROBLEM: **Given a raw score, find the corresponding frequency or percent frequency.**

1. Convert the raw score to a z score ($z = (X - \mu)/\sigma$).
2. Enter the normal curve table in the z column and read out the percent area from the mean to this z value.
3. Compute the answer to the problem. Note that the normal curve table gives only the area from the mean to the z value; be sure your final answer is the one called for by the problem.

2. TYPE 2 PROBLEM: **Given a frequency or percent frequency, find the corresponding raw score.**

1. Compute the percent frequency (if it is not given) and enter the normal curve table in the area portion. Read out the corresponding z score.
2. Convert the z score to a raw score ($X = z\sigma + \mu$).

Note that this is exactly the reverse of the procedure for Type 1 problems.

Always draw a rough normal curve diagram and fill in the appropriate numerical values. This will help clarify the problem and make it easier to arrive at the correct answer.

PROBLEMS

1. The following problems deal with the Scholastic Aptitude Test. Scores on this test may be assumed to be normally distributed in the population, and it is known that $\mu = 500$ and $\sigma = 100$.

a. What percent of the population obtains scores of 410 or less?

$z =$ ____________

Percent from table = ____________

Answer = ____________

b. What percent of the population obtains scores between 430 and 530?

$z =$ ____________

Percent from table = ____________

$z =$ ____________

Percent from table = ____________

Answer = ____________

c. What percent of the population obtains scores between 275 and 375?

$z =$ ____________

Percent from table = ____________

$z =$ ____________

Percent from table = ____________

Answer = ____________

d. How do you explain the fact that the answer to c is smaller than the answer to b, even though each problem deals with a 100-point range of scores?

e. What is the minimum score needed to rank in the top 5% of the population?

Answer = ____________

f. A psychologist wishes to conduct a learning experiment using those who perform very poorly on the SAT as subjects. More specifically, he wishes to select those in the bottom 15% of the population. What cutting score should he use?

Answer = ____________

g. A single student is drawn at random from the population. What is the probability that this student has a score:

of 410 or less? Answer = ______

between 430 and 530? Answer = ______

between 275 and 375? Answer = ______

in the top 5% of the population? Answer = ______

(Hint: no new calculations should be necessary.)

2. Susanne Trueblood is struggling diligently to understand the reasons for the normal curve procedure.

a. What one major problem in behavioral science research does the normal curve model enable us to solve? (Hint: the answer involves the size of the population.)

b. What assumption must be made when using the normal curve model?

c. What will happen if this assumption is incorrect?

d. Is the normal curve model the only statistical model used in inferential statistics?

e. How did the person who constructed the normal curve table obtain the percents within the table?

f. Why is the normal curve table given in terms of z scores, rather than raw scores?

ADDITIONAL COMPUTATIONAL PRACTICE

A population is normally distributed with $\mu = 63.7$ and $\sigma = 11.2$. Complete each of the tables that follow, drawing the appropriate diagram for each problem.

Table 1 *Type 1 problems*

percent of the population	*z value(s)*	*percent(s) from table*	*answer*
below 55	________	________	________
below 65	________	________	________
below 75	________	________	________
above 55	________	________	________
above 70	________	________	________
between μ and 53.2	________	________	________
between μ and 85.7	________	________	________
between 50 and 60	________	________	________
between 60 and 70	________	________	________
between 80 and 90	________	________	________

Table 2 *Type 2 problems*

raw score required to be in top:	*z value*	*answer*
$\frac{1}{2}$%	________	________
1%	________	________
$2\frac{1}{2}$%	________	________
5%	________	________
10%	________	________
25%	________	________
50%	________	________
60%	________	________
80%	________	________
90%	________	________
95%	________	________
$97\frac{1}{2}$%	________	________
99%	________	________
$99\frac{1}{2}$%	________	________

10. Inferences about the Mean of a Single Population

REMINDER 1. Hypothesis testing—general considerations

1. State the *null hypothesis* (denoted by the symbol H_0) and the *alternative hypothesis* (denoted by H_1). It is important to understand that you cannot *prove* whether H_0 or H_1 is true because you cannot measure the entire population.

2. Begin with the assumption that H_0 is true. Then obtain the data and test out this assumption. If this assumption is unlikely to be true, you will abandon it and "switch your bets" to H_1; otherwise, you will retain it.

   Before you collect any data, however, it is necessary to define in numerical terms what is meant by "unlikely to be true." In statistical terminology, this is called selecting a *criterion* (or *level*) *of significance*, represented by the symbol α.

   a. Using the .05 criterion of significance, "unlikely" is defined as having a probability of .05 or less. In symbols, $\alpha = .05$.

   b. Using the .01 criterion of significance. "unlikely" is defined as having a probability of .01 or less. In symbols, $\alpha = .01$.

   Other criteria of significance can be used, but these are the most common.

3. Having selected a criterion of significance, obtain your data and compute the appropriate statistical

test. If the test shows that H_0 is unlikely to be true, reject H_0 in favor of H_1. Otherwise, retain H_0.

4. Though you are backing the hypothesis indicated by the statistical test (H_0 or H_1), you could be wrong. In fact, you could make either of the following two kinds of error, depending on your decision:

   a. Type I Error: Reject H_0 when H_0 is in fact true. This error is less likely when the .01 criterion of significance is used; in fact, the probability of this kind of error (that is, the risk you run that your decision to reject H_0 is incorrect) is equal to the chosen criterion of significance α.

   b. Type II Error: Retain H_0 when H_0 is in fact false. This error is less likely when the .05 criterion of significance is chosen. The probability of this kind of error, denoted by the symbol β, is not so conveniently determined. (See Chapter 14.)

   Thus, the relative importance of each kind of error helps to determine what criterion of significance to use.

2. The standard error of the mean

The formula for estimating the standard error of the mean from a single sample is

$$s_{\bar{X}} = \frac{s}{\sqrt{N}}$$

where $s_{\bar{X}}$ = the estimated standard error of the mean

s = standard deviation of the sample of observations

Then, to draw inferences about the population mean, use

$$t = \frac{\bar{X} - \mu}{s_{\bar{X}}} \qquad \text{with} \quad df = N - 1$$

Use z instead of t if σ is known. For large samples, t and z are approximately equal.

3. The standard error of a proportion

If the data are in the form of proportions,

1. Compute

$$\sigma_p = \sqrt{\frac{\pi(1-\pi)}{N}}$$

2. Compute

$$z = \frac{p - \pi}{\sigma_p}$$

where σ_p = the standard error of a proportion
p = proportion observed in the sample
π = hypothesized value of the population proportion (*not* 3.14159)

Do *not* use this procedure if $N\pi$ or $N(1-\pi)$ is less than 5.

4. Confidence intervals

Use $s_{\bar{X}}$ when determining confidence intervals for the mean of one population. Obtain the critical value of t from the t table ($df = N - 1$; $\alpha = 1.0 -$ confidence) and compute

$$(\bar{X} - ts_{\bar{X}}) \leq \mu \leq (\bar{X} + ts_{\bar{X}})$$

where t = critical value from t table.

Use z instead of t if σ is known. For large samples, z and t are approximately equal.

PROBLEMS

1. Suppose that you have developed a new scientific theory which you hope will gain you lasting fame and fortune. Being a properly cautious scientist, you test out your theory by collecting data from a sample drawn randomly from the specified population and statistically analyzing the results.

a. The table below shows the possible outcomes of the statistical test and the possible true states of affairs. Fill in each of the cells with the proper designation: "Correct decision," "Type I error," or "Type II error."

		true state of affairs	
		H_0 *is true*	H_0 *is false*
your decision	*Retain* H_0		
	Reject H_0		

b. State the *practical* consequences of:

A Type I error:

A Type II error:

c. Which statistical model should be used if you wish to test hypotheses about:

1. proportions drawn from a single population?

2. the mean of a single population, where σ is *not* known?

2. State whether you agree or disagree with each of the following pearls of wisdom by Susanne Trueblood. If you disagree, give your reasons.

a. "If you are using the .05 criterion of significance, a result with a probability of .06 of occurring if H_0 is true is treated exactly the same as a result with a probability of .60 of occurring if H_0 is true."

b. "The probability of a Type I error when we use the .05 criterion is .05, and the probability of a Type I error when we use the .01 criterion is only .01. Since it is good to avoid a Type I error, we should always use the .01 criterion of significance."

c. "The probability of a Type I error is exactly equal to the criterion of significance, but the probability of a Type II error is not so conveniently determined."

d. "If we retain H_0 using the .05 criterion of significance, we may be 95% confident of being correct."

3. Believe it or not, Smedley Trueblood has come up with a useful idea for understanding the meaning of the standard error of the mean. (Perhaps this indicates that if a large enough sample of ideas is obtained, a small number may be significantly useful purely by chance.) This exercise should help to improve your understanding of this concept.

Make up a "population" as follows: Get 20 small slips of paper or file cards. On eight of the slips, write the number 50; on five of the slips, write the number 51; on five of the slips, write the number 49; on one of the slips, write the number 52; and on the last slip, write the number 48. Note that the mean of this population is equal to 50. Place all the slips in a hat or container and mix thoroughly.

For purposes of this problem, we will assume that it is impossible to measure the entire population and that we must draw a sample in order to estimate the population mean.

a. Randomly draw one slip of paper from the container and record the number in the first space below. Replace the slip of paper in the container, shuffle thoroughly, and draw a second slip of paper. Record the number in the second space below. Repeat this process until five observations have been drawn and the numbers recorded. Then, compute the mean of the five numbers.

______ ______ ______ ______ ______ $\bar{X}_1 =$ ______

b. Repeat the procedure in part a.

______ ______ ______ ______ ______ $\bar{X}_2 =$ ______

c. Again, repeat the procedure in part a.

________ ________ ________ ________ ________ $\bar{X}_3 =$ ________

d. Repeat the procedure in part a once again.

________ ________ ________ ________ ________ $\bar{X}_4 =$ ________

e. Once more, repeat the procedure in part a.

________ ________ ________ ________ ________ $\bar{X}_5 =$ ________

f. One last time, repeat the procedure in part a.

________ ________ ________ ________ ________ $\bar{X}_6 =$ ________

g. Unless you have been the victim of a most improbable occurrence, the six means above vary. Now, compute the standard deviation of the six means, using $N-1$ in the denominator. (Do *not* use the formula $s/\sqrt{N}$ when you are done—you are computing the standard error of the mean directly.)

s of these six means $= s_{\bar{X}} =$ ________

The $s_{\bar{X}}$ that you have just computed is an estimate of the standard error of the mean. (The more means, the better the estimate.) It indicates how much one sample mean is likely to vary from the next due to sampling error—those cases which, by chance, happened to be included in the sample.

h. Clearly, drawing all these samples is a lot of work; and in research where considerably more effort than drawing numbers out of a hat is involved, it is far too expensive and time-consuming. (If many cases are available, it is best statistically to treat them as one large sample.) Therefore, procedures are used for *estimating* $s_{\bar{X}}$ *from a single sample*.

For example, suppose that only one sample had been drawn—say the sample in a above. Compute the standard deviation of this sample; then compute the estimated standard error of the mean using the formula $s/\sqrt{N}$. Compare the result to that obtained in part g, noting that they may well be somewhat different because of the small size of the sample.

s of five raw scores in a = ________

$s_{\bar{X}} = s/\sqrt{N} =$ ________

4. Now make up 20 new slips of paper. Use the following numbers, one on each slip:

5 10 15 20 25 30 35 40 45 50

50 55 60 65 70 75 80 85 90 95

(It is unlikely that a population would be distributed in this manner, but anything is fair in an illustrative example.) Note that the population mean is equal to 50. Repeat steps a through g of problem 3 of this chapter for this new population.

a. ______ ______ ______ ______ ______ $\bar{X}_1 =$ ______

b. ______ ______ ______ ______ ______ $\bar{X}_2 =$ ______

c. ______ ______ ______ ______ ______ $\bar{X}_3 =$ ______

d. ______ ______ ______ ______ ______ $\bar{X}_4 =$ ______

e. ______ ______ ______ ______ ______ $\bar{X}_5 =$ ______

f. ______ ______ ______ ______ ______ $\bar{X}_6 =$ ______

g.

s of these six means $= s_{\bar{X}} =$ ______

While you may have been the victim of sampling error and chance, it is overwhelmingly probable that you found this second standard error of the mean to be larger than the one for the population in problem 3.

h. In which case (problem 3 or this problem) can we be more confident in any one sample mean as an estimate of the population mean?

i. Suppose you were to repeat step h of problem 3 for the data in this problem. Without computing anything, what would happen to the size of s and the size of $s/\sqrt{N}$?

j. If the size of the sample is increased, what happens to the size of $s_{\bar{X}}$? Does this imply that larger samples are more desirable?

5. a. What conditions must be satisfied for a sample to be considered randomly drawn?

b. Which of the following are random samples from the specified populations and which are not? Why?

1. An experimenter selects every tenth name, starting with one chosen blindly, in the Manhattan telephone book. (Population: All telephone owners in Manhattan with listed numbers.)

2. Students in two 8 A.M. classes in psychology at a large university are selected from the population of all students at the university to participate in an experiment.

3. Students sitting in the first three rows of a large class are selected for one condition of an experiment, students in the middle three rows take part in a second aspect of the experiment, and students in the back rows are used for a third part of the experiment. (Population: All students who will ever take the course in question.)

4. The names of all students at a large state university are written on slips of paper and placed in an extremely large hat. The slips are well shuffled and 50 names are drawn blindly to serve as subjects in an experiment. (Population: All students enrolled in this university.)

5. An experimenter running rats through a maze reaches into a cage with his eyes closed and picks the first ten rats he gets his hands on. (Population: All rats in this cage.)

6. For each of the following problems, first compute $s_{\bar{X}}$; then answer the question using the .05 criterion of significance.

a. Would you retain or reject the null hypothesis that the population mean for University A is 14? (The university means and standard deviations were computed in Chapter 4, problem 1, and Chapter 5, problem 3.)

$s_{\bar{X}} =$ ______

$t =$ ______

Decision: ______

b. Would you retain or reject the null hypothesis that the population mean for University B is 10?

$s_{\bar{X}} =$ ....

$t =$ ....

Decision: ....

c. Students at Bigbrain University claim that they are superior to the general run-of-the-mill college student because they have found that for a random sample of 25 students, the mean Scholastic Aptitude Test Score is 530 and the standard deviation is 100. Is their claim justified, or not?

$s_{\bar{X}} =$ ....

$t =$ ....

Decision about H_0: ....

Decision about claim: ....

d. A group of students at Old Siwash U. who are well trained in statistics makes a similar claim regarding their school's superiority. They have found that for a random sample of 64 students, the SAT mean is 530 and the standard deviation is 100. Is their claim justified, or not?

$s_{\bar{X}} =$ ____________

$t =$ ____________

Decision about H_0: ____________

Decision about claim: ____________

7. A politician has staked his political career on whether or not a new state constitution will pass. To find out which way the wind is running, he obtains a random sample of 100 voters a few weeks prior to the election and finds that 60% of the sample will vote for the new constitution. Assuming that the constitution will fail if it receives 50% or less of the vote, should he conclude that the electorate as a whole will support the constitution?

8. Aside from his general confusion concerning statistics, Smedley is depressed because of the following homework assignment, which he thinks will involve a great deal of computational effort:

"Test the null hypothesis that the University A population mean is

a. 11.0
b. 11.3
c. 11.7
d. 12.2
e. 12.8
f. 13.4
g. 13.7
h. 14.0
i. 15.0

(Use the .05 criterion of significance in all of the above tests.)"

Smedley grimly sets out to test each of the above hypotheses, using the standard error of the mean procedures. After doing this nine times, he begins to wonder if there is not an easier way to do it. Compute the 95% confidence interval for the University A data and show how this makes it possible to test all of the above hypothesized values of the mean of population A.

9. For University B, compute each of the following:

a. The 99% confidence interval.

b. The 95% confidence interval.

c. The 90% confidence interval.

d. Explain why the size of the confidence interval becomes larger as we go from 90% to 99% confidence intervals.

10. a. In another of her typical bursts of inspiration, Susanne Trueblood argues that we should always use 99.99% confidence intervals, because these intervals are larger than the 95% or 99% confidence intervals and we can therefore be more confident that the population mean falls within them. Do you agree or disagree? Why?

b. How does the size of the confidence interval depend on the size of the standard error of the mean?

ADDITIONAL COMPUTATIONAL PRACTICE

Assume that each of the data sets named below is a random sample from a large population. For each set, test the indicated null hypothesis using the .05 criterion of significance; then compute the indicated confidence intervals. (For convenience, first write down the previously computed mean, s, and N for each set.)

data set 1, chapter 4

$\bar{X} =$ ________

$s =$ ________

$N =$ ________

Hypothesis testing:

H_0: $\mu = 2.5$

H_1: $\mu \neq 2.5$

$s_{\bar{X}} =$ ________

$t =$ ________

Decision about H_0: ________

Confidence intervals:

95%: ________ $\leq \mu \leq$ ________

99%: ________ $\leq \mu \leq$ ________

data set 2, chapter 4

$\bar{W} =$ ________

$s =$ ________

$N =$ ________

Hypothesis testing:

H_0: $\mu = 52.3$

H_1: $\mu \neq 52.3$

$s_{\bar{X}} =$ ________

$t =$ ________

Decision about H_0: ________

Confidence intervals:

95%: ________ $\leq \mu \leq$ ________

99%: ________ $\leq \mu \leq$ ________

data set 3, chapter 4

$\bar{X} =$ ________

$s =$ ________

$N =$ ________

Hypothesis testing:

H_0: $\mu = 6.0$

H_1: $\mu \neq 6.0$

$s_{\bar{X}} =$ ________

$t =$ ________

Decision about H_0: ________

Confidence intervals:

95%: ________ $\leq \mu \leq$ ________

99%: ________ $\leq \mu \leq$ ________

Y data from set 3, chapter 1

$\bar{Y} =$ ________

$s =$ ________

$N =$ ________

Hypothesis testing:

H_0: $\mu = 85.0$

H_1: $\mu \neq 85.0$

$s_{\bar{X}} =$ ________

$t =$ ________

Decision about H_0: ________

Confidence intervals:

95%: ________ $\leq \mu \leq$ ________

99%: ________ $\leq \mu \leq$ ________

A certain business concern needs to obtain at least 20% of the market in order to make a profit. A random sample of 200 prospective buyers is asked whether or not it will purchase the product. What should the company conclude in each of the following cases?

result: percent of sample that will buy product	σ_p	z	*decision about H_0*	*decision about profit*
13%	________	________	________	________
23%	________	________	________	________
29%	________	________	________	________

11. Testing Hypotheses about Differences between the Means of Two Populations

REMINDER 1. **For independent samples,** first compute

$$s^2_{\text{pooled}} = \frac{(N_1 - 1)s_1{}^2 + (N_2 - 1)s_2{}^2}{N_1 + N_2 - 2}.$$

Then use

$$t = \frac{\bar{X}_1 - \bar{X}_2}{s_{\bar{X}_1 - \bar{X}_2}} = \frac{\bar{X}_1 - \bar{X}_2}{\sqrt{s^2_{\text{pooled}}\left(\frac{1}{N_1} + \frac{1}{N_2}\right)}}$$

with $df = N_1 + N_2 - 2$

where $s_{\bar{X}_1 - \bar{X}_2}$ = the estimated standard error of the difference

Note that this t formula may well yield misleading results if σ_1^2 and σ_2^2 are markedly unequal *and* also N_1 and N_2 are markedly unequal.

Note: The two-step procedure shown above is sometimes expressed in the following form, which yields exactly the same result:

$$t = \frac{\bar{X}_1 - \bar{X}_2}{\sqrt{\dfrac{\sum X_1^2 + \sum X_2^2 - \dfrac{(\sum X_1)^2}{N_1} - \dfrac{(\sum X_2)^2}{N_2}}{N_1 + N_2 - 2}\left(\dfrac{1}{N_1} + \dfrac{1}{N_2}\right)}}$$

with $df = N_1 + N_2 - 2$

2. **For matched samples,** first compute a difference score (symbolized by D) for each pair, where

$$D = X_1 - X_2$$

Next, compute the *variance* of the D scores, using

the usual formula for the variance of a sample (Chapter 5). Then use

$$t = \frac{\bar{D}}{\sqrt{\frac{s_D^2}{N}}}$$

with $df =$ (Number of pairs) $- 1$

where $\bar{D} =$ mean of the D scores
$s_D^2 =$ variance of the D scores
$N =$ number of pairs

Remember to keep all signs when computing D.

Note: The procedure shown above may be summarized in the following computing formula, which yields exactly the same result:

$$t = \frac{\sum D}{\sqrt{\frac{N \sum D^2 - (\sum D)^2}{N-1}}}$$

with $df =$ (number of pairs) $- 1$

PROBLEMS

1. In recognition of Smedley Trueblood's one good idea in four years, let us use the results of Chapter 10, problems 3 and 4, to illustrate what is meant by the standard error of the difference. List the six sample means drawn from population 1 (problem 3) and the six sample means drawn from population 2 (problem 4) in the table below, and then compute the difference between each pair of means.

	population 1 ($\bar{X}_1$)	*population 2* ($\bar{X}_2$)	$\bar{X}_1 - \bar{X}_2$
Sample (a)	....	....	....
Sample (b)	....	....	....
Sample (c)	....	....	....
Sample (d)	....	....	....
Sample (e)	....	....	....
Sample (f)	....	....	....

Each of the entries in the $\bar{X}_1 - \bar{X}_2$ column is an estimate of the difference between the population means. (In this experiment, we know that the population means are equal; but remember that we are assuming that we do not know the values of the population means and must draw inferences about them based on the sample means, as would be the case in a real study.) Now compute the standard deviation of the six differences in the $\bar{X}_1 - \bar{X}_2$ column, using $N - 1$ in the denominator.

s of these six differences = $s_{\bar{X}_1 - \bar{X}_2}$ = ____________

This standard deviation is an estimate of the standard error of the difference. (The more differences, the better the estimate.) It is a measure of how much the *difference* between two sample means is likely to vary from trial to trial. Therefore, it gives an indication as to how much confidence can be placed in a single difference of a given size as an estimate of the difference between the population means.

In practice, it is too expensive and time-consuming to draw a large number of samples. Fortunately, the standard error of the difference can be *estimated using the data from a single sample*, using the procedure shown in the reminder for this chapter.

2. The students of Universities A and B have placed a bet regarding which population is superior on the psychology quiz. If the difference between the sample means is not statistically significant, the bets are withdrawn and no one wins. Test the difference for statistical significance. What is your conclusion?

$t =$ ....

Decision about H_0: ....

Decision about bet: ....

3. An industrial psychologist obtains scores on a test used to help select people for jobs from 41 men and 31 women. The results are: men–mean = 48.75 and standard deviation = 9.0; women–mean = 46.07 and standard deviation = 10.0. Test the difference between men and women for statistical significance. What should the psychologist decide?

$t =$ ....

Decision about H_0: ....

Decision about men and women: ....

4. The same psychologist as in problem 3 of this chapter also wishes to investigate whether people of different ethnic backgrounds score at different levels on tests used to select people for jobs. For the data shown below, test the significance of the difference. What should the psychologist decide?

ethnic group 1 (X)	*ethnic group 2* (Y)
62	46
54	53
59	50
56	52
59	54

$\bar{X} =$ ____________

$\bar{Y} =$ ____________

$s_X^2 =$ ____________

$s_Y^2 =$ ____________

$t =$ ____________

Decision about H_0: ____________

Decision about ethnic groups: ____________

5. An educational psychologist has developed a new textbook based on programmed instruction techniques and wishes to know if it is superior to the conventional kind of textbook. He therefore obtains subjects who have had no prior exposure to the material and forms two groups: an experimental group, which learns via the programmed text, and a control group, which learns via the old-fashioned textbook.

The psychologist is afraid, however, that one group may differ from the other in intelligence. If such is the case, differences in the effectiveness of one or the other of the learning procedures may be obscured by differences in ability to learn. Therefore, he matches his subjects on intelligence, and forms 10 pairs of subjects such that each pair is made up of two people roughly equal in intelligence test scores.

After both groups have learned the material, the psychologist measures the amount of learning by means of a 10-item quiz. The results are as follows:

pair	*experimental group (programmed text)*	*control group (standard text)*
1	9	7
2	6	4
3	5	6
4	7	3
5	3	5
6	7	3
7	3	2
8	4	5
9	6	7
10	10	8

Test the results for statistical significance. What should the psychologist decide about his new programmed text?

$t =$ ....

Decision about H_0: ....

Decision about programmed text: ....

6. a. In problem 5, which type of error might you have made, Type I or Type II? What would be the practical consequences of such an error?

b. In problem 2, which type of error might you have made, Type I or Type II? What would be the practical consequences of such an error?

c. What is the probability of making the type of error discussed in part b?

d. Why must we risk making either a Type I error or Type II error in experiments like these?

e. What one major problem does our statistical model, the *t* distributions, enable us to solve in experiments like these?

7. a. Susanne Trueblood argues that if the results of the statistical test indicate that H_0 should be retained, we have proved that there is no difference between the population means. Do you agree? Why or why not?

b. "The probability of a Type I error when using the t test for the difference between two means is equal to the criterion of significance but the probability of a Type II error is not so conveniently determined." Is this statement true?

c. Susanne wants to do a research study involving the t test for the difference between two means, but is afraid that the raw scores in each population are not normally distributed. What advice would you give her?

d. In another plan for a study involving the t test for the difference between two means, Susanne has reason to believe that the variance of one population is much different from the variance of the other. What should she do?

ADDITIONAL COMPUTATIONAL PRACTICE

For each of the two (separate) sets of data shown below, test the difference between the means for statistical significance.

data set 1

	group 1	group 2
$\bar{X}$	17.34	21.58
s	5.83	4.42
N	32	30

$s^2_{\text{pooled}} =$ ____________________

$t =$ ____________________

Decision about H_0: ____________________

data set 2

	group 1	group 2
$\bar{X}$	76.57	69.72
s	20.15	22.87
N	17	15

$s^2_{\text{pooled}} =$ ____________________

$t =$ ____________________

Decision about H_0: ____________________

For the set of data below, assume that the X score represents the subject's performance in the experimental condition and that the Y score represents the subject's performance in the control condition. (Thus, each subject serves as his or her own control.) Compute the *matched* t test for these data.

data set 3, chapter 1

S	X	Y	D
1	97	89	______
2	68	57	______
3	85	87	______
4	74	76	______
5	92	97	______
6	92	79	______
7	100	91	______
8	63	50	______
9	85	85	______
10	87	84	______
11	81	91	______
12	93	91	______
13	77	75	______
14	82	77	______

$t =$ ______

Decision about H_0: ______

Now carry out a matched t test for the set of data shown below.

same data, but with X rearranged

S	X	Y	D
1	92	89	________
2	82	57	________
3	85	87	________
4	81	76	________
5	87	97	________
6	93	79	________
7	68	91	________
8	85	50	________
9	63	85	________
10	100	84	________
11	74	91	________
12	77	91	________
13	92	75	________
14	97	77	________

$t =$ ____________

Decision about H_0: ____________

12. Linear Correlation and Prediction

REMINDER 1. The Pearson correlation coefficient

Z score difference formula:

$$r_{XY} = 1 - \frac{1}{2}\frac{\sum (Z_X - Z_Y)^2}{N}$$

raw score definition formula:

$$r_{XY} = \frac{\sum (X - \bar{X})(Y - \bar{Y})}{N\sigma_X \sigma_Y}$$

Z score product formula:

$$r_{XY} = \frac{\sum Z_X Z_Y}{N}$$

raw score computing formula:

$$r_{XY} = \frac{N \sum XY - \sum X \sum Y}{\sqrt{[N \sum X^2 - (\sum X)^2][N \sum Y^2 - (\sum Y)^2]}}$$

In the above formulas, $N =$ number of *pairs*.

REMEMBER

a. The numerical value of r indicates the *strength* of the relationship between the two variables, while the sign indicates the *direction* of the relationship.

b. $-1 \leq r \leq +1$.

c. r *cannot* be simply interpreted as a percent, but r^2 can.

d. r may be tested for statistical significance by referring the obtained value to the appropriate table with $df = N - 2$.

e. r may also be used as a descriptive statistic.

f. r detects only *linear* relationships.

g. Correlation does not imply causation.

The population correlation is symbolized by ρ.

2. Linear regression

The goal is to obtain the predicted score Y', given a score on X. The formula to use is:

$$Y' = b_{YX}X + a_{YX}$$

where

$$b_{YX} = r_{XY}\frac{\sigma_Y}{\sigma_X} = \frac{N\sum XY - \sum X \sum Y}{N\sum X^2 - (\sum X)^2}$$

$$a_{YX} = \bar{Y} - b_{YX}\bar{X}$$

The standard error of estimate for predicting Y from X, a measure of error in the prediction process, is equal to

$$\sigma_{Y'} = \sqrt{\frac{\sum(Y - Y')^2}{N}} = \sigma_Y\sqrt{1 - r_{XY}^2}$$

This measure is a descriptive statistic. If instead you wish to estimate the value of the corresponding population parameter, compute

$$s_{Y'} = \sqrt{\frac{\sum(Y - Y')^2}{N-2}}$$

REMEMBER

a. The above procedure minimizes the sum of squared errors in prediction, $\sum(Y - Y')^2$.
b. The closer r is to zero, the greater the errors in prediction. If $r = \pm 1$, $\sigma_{Y'} = 0$ (prediction is perfect). If $r = 0$, $\sigma_{Y'}$ equals its maximum possible value σ_Y (prediction is useless).
c. If r is not statistically significant, linear regression should *not* be used.
d. The regression line always passes through the point $\bar{X}$, $\bar{Y}$.
e. The above procedure is used only for purposes of predicting scores on Y (Y is the criterion and X is the predictor). Different formulas are needed to predict scores on X, and the simplest procedure in such cases is to change the designations (relabel X as Y and Y as X) so that the above procedure may be used.

PROBLEMS

1. A college dean would like to know how well he can predict sophomore grade point average for first-semester freshmen, so that students who are headed for trouble can be given appropriate counseling. After students have been in school for one semester, the dean obtains their numerical final examination average for the first semester (based on a total of 100 points) and the average number of "cuts" per class during the semester. He then waits for a year and a half, and when the students have finished their second year he obtains their sophomore grade point average.

A large sample would be desirable for such a study. Since the purpose of this problem is to see how correlation and regression procedures work, let us keep the computations within reason by assuming that the dean has a sample of only seven cases, remembering that in a real study there would be many more subjects (but exactly the same procedures would be used).

student	*test score* (X)	*cuts* (C)	*sophomore average* (Y)
1	70	2	2.50
2	90	1	4.00
3	75	2	3.50
4	85	3	3.00
5	80	5	3.00
6	70	3	2.00
7	90	5	3.00
Mean	80	3	3.00
σ	8.02	1.41	.60

a. Convert the test scores (X) and sophomore averages (Y) to Z scores. By inspection of the paired Z scores, estimate whether the correlation between these two variables is strong and positive, about zero, or strong and negative; then verify your estimate by computing r using the *Z score difference formula*.

student	Z_X	Z_Y	$Z_X - Z_Y$	$(Z_X - Z_Y)^2$
1				
2				
3				
4				
5				
6				
7				

$\sum (Z_X - Z_Y)^2 =$ ....

$r_{XY} =$ ....

b. Now recompute r_{XY} by the *Z score product formula* and verify that the result is the same. Briefly, state in words what this r_{XY} indicates.

$\sum Z_X Z_Y =$ ....

$r_{XY} =$ ....

c. Compute r_{CY} using the *raw score computing formula*. State briefly what this correlation coefficient indicates.

$\sum C =$ ____________

$\sum C^2 =$ ____________

$\sum Y =$ ____________

$\sum Y^2 =$ ____________

$\sum CY =$ ____________

$N \sum CY - \sum C \sum Y =$ ____________

$N \sum C^2 - (\sum C)^2 =$ ____________

$N \sum Y^2 - (\sum Y)^2 =$ ____________

$r_{CY} =$ ____________

d. Recompute r_{CY} using the *raw score definition formula* and verify that the result is the same.

student	$C - \bar{C}$	$Y - \bar{Y}$	$(C - \bar{C})(Y - \bar{Y})$
1			
2			
3			
4			
5			
6			
7			

$\sum (C - \bar{C})(Y - \bar{Y}) =$ ____________

$N\sigma_C \, \sigma_Y =$ ____________

$r_{CY} =$ ____________

i. Is it desirable that $\sum (Y - Y')^2$, and hence $\sigma^2_{Y \cdot}$ be small or large? Why?

j. What are the smallest and largest possible values of $\sigma_{Y'}$?

k. Using the result of part e, determine what the predicted sophomore average would be for a new student with a score of 72.

l. Is more weight given to scores on X in the equation for predicting scores on Y when r_{XY} is large in absolute value or when it is small? Why?

2. a. A psychologist wishes to know if a correlation exists between the personality traits of sociability and cheerfulness. She obtains a random sample of 32 subjects and administers measures of both traits. She finds that the sample correlation coefficient is +.38. What should the psychologist decide, and why?

b. An industrial psychologist has designed a test to predict whether or not workers will be successful on assembly-line jobs. A random sample of 47 job applicants are given a test. Regardless of test scores, all applicants are hired, and six months later a measure of job proficiency is obtained. The correlation between the test and the job proficiency measure is found to be +.18. Explain what the psychologist should decide in this case and why, *and* explain why applicants with poor scores on this test should nevertheless *not* be considered "poor risks."

3. A psychologist hypothesizes that psychotherapists who are more outgoing will be more successful in avoiding missed appointments by their patients. The psychologist obtains a random sample of 30 therapists and administers an introversion–extraversion scale to each one, where *high* numerical scores reflect greater extraversion (more outgoing) and *low* scores reflect greater introversion (more shy). The psychologist also obtains the number of appointments per week missed by the patients of each therapist. The correlation between introversion–extraversion and appointments missed is +.11.

a. Is the *direction* of this co-relationship in agreement with or opposite to the psychologist's hypothesis? Explain your answer.

b. Test this result for statistical significance. What is your decision about H_0? About the psychologist's hypothesis?

c. Which type of error might you have made in part b, Type I or Type II? What would be the practical consequences of such an error?

d. Now suppose instead that the correlation between introversion–extraversion and appointments missed is −.43. Test this result for statistical significance. What is your decision about H_0? About the psychologist's hypothesis?

e. Which type of error might you have made in part d, Type I or Type II? What would be the practical consequences of such an error?

f. Given the result in part d, is it possible to determine which variable (introversion–extraversion or appointments missed) is the cause and which is the effect? Justify your answer by showing either that it is logical or impossible for either of the two variables to be the cause or the effect.

4. a. Smedley Trueblood is highly excited about an idea for a research study. He plans to correlate scores on this year's statistics final examination with scores on last year's statistics final to see if test performance is consistent from year to year. What is your evaluation of this research plan?

b. Smedley just happens to have obtained identical scores of 72 on his English midterm and 72 on his history midterm. Based on this information, he seeks to impress his professors by telling them that the correlation between the English and history tests is +1.00. Is he right?

c. If the correlation between the statistics midterm examination and the statistics final examination is $+.65$ and Smedley (believe it or not) is above average on the midterm, what can be said about his score on the final?

d. If the correlation between number of hours spent studying and number of mistakes on the final examination in statistics is $-.25$ and Smedley is above average in hours studying, what can be said about his performance on the final? Is this prediction or the one in the preceding question more likely to be correct? Why?

e. Smedley argues that the correlation of —.25 in problem d is very illogical. He claims that since studying more usually improves test performance, the correlation should be positive. Is he correct?

f. In a research study, Smedley finds that $r_{XY} = .06$ for $df = 2000$, which is statistically significant at the .05 level. Smedley therefore concludes that a strong positive relationship exists between X and Y in the population. Do you agree or disagree, and why?

g. Smedley is confused by linear regression and argues as follows: "The purpose of linear regression is to obtain predicted scores Y'. However, we must know the actual scores on Y to be able to determine the linear regression equation. Since there is no point in predicting what we already know, linear regression is totally useless." Explain what is wrong with his reasoning.

h. Suppose that the dean in problem 1 of this chapter had happened to wind up with a sample of seven honor students, all of whom achieved sophomore grade point averages of 3.30 or more, and that he proceeded with the statistical analysis without realizing that the sample was out of the ordinary. What would be the probable effect on the correlation between test scores and sophomore grade-point average, and on the dean's conclusion as to the merits of using the test scores for prediction?

ADDITIONAL COMPUTATIONAL PRACTICE

For each of the sets of data that follow, compute the values needed in order to fill in the answer spaces. (Some of these values will already have been computed in preceding chapters.)

data set 3, chapter 1

S	X	Y
1	97	89
2	68	57
3	85	87
4	74	76
5	92	97
6	92	79
7	100	91
8	63	50
9	85	85
10	87	84
11	81	91
12	93	91
13	77	75
14	82	77

$\sum X =$ ______ $\sum Y =$ ______

$\sum X^2 =$ ______ $\sum Y^2 =$ ______

$\sum XY =$ ______ $N =$ ______

$(\sum X)^2 =$ ______ $(\sum Y)^2 =$ ______

$\bar{X} =$ ______ $\bar{Y} =$ ______

$\sigma_X =$ ______ $\sigma_Y =$ ______

$r_{XY} =$ ______

$b_{YX} =$ ______

$a_{YX} =$ ______

$\sigma_{Y'} =$ ______

X score	*predicted Y′ score*
96	______
92	______
90	______
85	______
80	______
70	______

same data, but with X rearranged

S	X	Y
1	92	89
2	82	57
3	85	87
4	81	76
5	87	97
6	93	79
7	68	91
8	85	50
9	63	85
10	100	84
11	74	91
12	77	91
13	92	75
14	97	77

$\sum XY =$ ....

$r_{XY} =$ ....

$b_{YX} =$ ....

$a_{YX} =$ ....

$\sigma_{Y'} =$ ....

X score	*predicted Y′ score*
96	....
92	....
90	....
85	....
80	....
70	....

(Except for computational practice, should Y' values be computed in this situation?)

same data, but with
X rearranged (again)

S	X	Y
1	82	89
2	92	57
3	81	87
4	85	76
5	68	97
6	93	79
7	77	91
8	100	50
9	87	85
10	85	84
11	63	91
12	74	91
13	97	75
14	92	77

$\sum XY =$ ....

$r_{XY} =$ ....

$b_{YX} =$ ....

$a_{YX} =$ ....

$\sigma_{Y'} =$ ....

X score	*predicted Y′ score*
96	....
92	....
90	....
85	....
80	....
70	....

13. Other Correlational Techniques

REMINDER 1. The Spearman rank-order correlation coefficient

With ranked data, use the Spearman rank-order correlation coefficient:

$$r_s = 1 - \frac{6 \sum D^2}{N(N^2 - 1)}$$

where D = difference between a pair of ranks
N = number of pairs

When $N < 10$, r_s must be referred to a different table than the Pearson r in order to determine statistical significance. For $N \geq 10$, the procedure used with the Pearson r gives a good approximation.

2. The point biserial correlation coefficient

With one continuous and one dichotomous variable, use the point biserial correlation coefficient:

$$r_{pb} = \frac{\bar{Y}_1 - \bar{Y}_0}{\sigma_Y} \sqrt{pq}$$

where X = dichotomized variable (equals 1 or 0)

Y = continuous variable

p = proportion scored 1 on the dichotomous variable

q = proportion scored 0 on the dichotomous variable

$\bar{Y}_1$ = mean score on Y for those scored 1 on X

$\bar{Y}_0$ = mean score on Y for those scored 0 on X

σ_Y = standard deviation of Y scores

Alternatively, the correlation could be computed by using the formula for the Pearson r (Chapter 12). r_{pb} is tested for significance in the same way as the Pearson r.

3. Converting significant values of t to r_{pb}

When a statistically significant value of t in a test of the difference between means is obtained, it is very desirable to convert it to r_{pb} so as to determine the *strength* of the relationship. This is readily done as follows:

$$r_{pb} = \sqrt{\frac{t^2}{t^2 + df}}$$

where $df = N_1 + N_2 - 2$

PROBLEMS

1. Ten subjects participate in a problem-solving experiment. Two judges are asked to rank order the solutions with regard to their creativity (1 = most creative, 10 = least creative). The experimenter wishes to

know if the judges are in substantial agreement. The rankings are shown below; what should the experimenter decide?

subject	*judge 1*	*judge 2*
1	5	4
2	7	9
3	2	2
4	9	8
5	1	3
6	4	1
7	10	10
8	3	7
9	6	5
10	8	6

$\sum D^2$ = ----------------

r_s = ----------------

Decision about H_0: ----------------

Decision about judges: ----------------

2. An industrial psychologist asks a group of 9 assembly-line workers and 11 workers not on an assembly line (but doing similar work) to indicate how much they like their jobs on a nine-point scale (9 = like, 1 = dislike). The results are as follows:

Assembly-line workers:	4 4 4 2 1 5 3 4 3
Other workers:	6 5 7 5 3 7 6 8 7 3 3

Test the null hypothesis that there is no relationship between the assembly line variable and job satisfaction by computing r_{pb}. What should the psychologist decide?

$\bar{Y}_1$ = ---------

$\bar{Y}_0$ = ---------

σ_Y = ---------

$\sqrt{pq}$ = ---------

r_{pb} = ---------

Decision about H_0: -----------------

Decision about relationship: -----------------

----

3. a. Suppose that the psychologist in problem 2 of this chapter had instead performed a t test for the difference between the mean satisfaction of assembly-line workers and the mean satisfaction of other workers. Without computing anything, would the resulting t value be statistically significant? Would this t value be more or less informative than r_{pb}?

b. Smedley Trueblood has just computed the t test for the difference between two means and has obtained a statistically significant value of 2.0; the sizes of the two samples are 75 and 67. He exultantly proclaims that he has at last produced a finding of great practical importance because it is statistically significant. Convert his t value to a point biserial correlation coefficient. Is he correct?

c. Suppose instead that Smedley had obtained a significant value of t of 3.0 with sample sizes of 15 and 14. What would your conclusion be in this case?

d. Convert any statistically significant values of t obtained in Chapter 11 of this workbook to values of r_{pb}. Briefly discuss the implications of each result.

e. Given a particular statistically significant value of t, is the relationship between the two variables stronger in the sample if it is based on large sample sizes or small sample sizes? Explain briefly why this makes good sense logically.

ADDITIONAL COMPUTATIONAL PRACTICE

For each of the two (separate) sets of data that follow, compute the Spearman rank-order correlation coefficient and test it for statistical significance.

data set 1

S	*rank by judge 1*	*rank by judge 2*	*D*
1	16	17	______
2	4	7.5	______
3	16	16	______
4	2	5	______
5	9	13	______
6	11.5	11	______
7	16	15	______
8	1	3	______
9	11.5	11	______
10	6.5	2	______
11	14	11	______
12	11.5	9	______
13	4	1	______
14	6.5	14	______
15	11.5	7.5	______
16	4	4	______
17	8	6	______

$\sum D^2 =$ ______

$r_s =$ ______

data set 1 with ranks of judge 2 rearranged

S	*rank by judge 1*	*rank by judge 2*	*D*
1	16	3	________
2	4	11	________
3	16	2	________
4	2	1	________
5	9	11	________
6	11.5	17	________
7	16	6	________
8	1	15	________
9	11.5	9	________
10	6.5	13	________
11	14	16	________
12	11.5	14	________
13	4	5	________
14	6.5	7.5	________
15	11.5	11	________
16	4	4	________
17	8	7.5	________

$\sum D^2 =$ ________

$r_s =$ ________

Convert each of the following statistically significant values of t, obtained from the t test for the difference between two independent means, to r_{pb}.

t	N_1	N_2	r_{pb}
2.11	12	8	________
2.75	12	8	________
6.00	12	8	________
2.11	19	23	________
2.75	19	23	________
6.00	19	23	________
2.11	51	51	________
2.75	51	51	________
6.00	51	51	________

14. Introduction to Power Analysis

REMINDER 1. The meaning of power

Power is the *probability of getting a significant result* in a statistical test. Power is equal to $1 - \beta$, where $\beta =$ the probability of a Type II error.

If power is not known, a researcher may waste a great deal of time by conducting an experiment that has little chance to produce significance even if H_0 is false. Worse, a promising line of research may be abandoned because the researcher does not know that he should have relatively little confidence concerning his failure to reject H_0 (that is, the probability of a Type II error is large).

2. The four major parameters of power analysis

1. *The significance criterion,* α. A Type II error is more likely as α gets smaller because you fail to reject H_0 more often. Thus, power decreases as α decreases.

2. *The sample size, N.* Larger samples yield better estimates of population parameters and make it more likely that you will reject H_0 when it is correct to do so. Thus, power increases as N increases.

3. *The population "effect" size,* γ *(gamma).* You are less likely to fail to reject a false H_0 if H_0 is "very wrong"—that is, if there is a large effect size in the population. Thus, power increases as γ increases.

4. *Power,* $1 - \beta$.

Any one of these four parameters is an exact mathematical function of the other three.

3. The general procedure for the two most important kinds of power analysis

1. Power determination
   a. Compute or posit the value of γ.
   b. Compute δ (delta), which combines N and γ.
   c. Obtain power from the appropriate table.
2. Sample size determination
   a. Specify desired power. (If a conventional value is needed, use .80.)
   b. Compute or posit the value of γ.
   c. Obtain δ from the appropriate table.
   d. Compute N.

If you are unable to specify the population values needed to compute γ, use the appropriate conventional value of γ for the statistical test in question.

4. The specific formulas for γ, δ, and N depend on the statistical test that is being performed.

1. The test of the mean of a single population

   a. $\gamma = \dfrac{\mu_1 - \mu_0}{\sigma}$

   where μ_0 = value of μ specified by H_0
   μ_1 = value of μ specified by H_1 (A *specific* H_1 is always necessary in power analysis)
   σ = population standard deviation

   b. $\delta = \gamma\sqrt{N}$

   c. $N = \left(\dfrac{\delta}{\gamma}\right)^2$

   d. Conventional values of γ: "small," .20; "medium," .50; "large," .80

2. The test of the proportion of a single population

   a. $\gamma = \dfrac{\pi_1 - \pi_0}{\sqrt{\pi_0(1 - \pi_0)}}$

   where π_0 = value of π specified by H_0
   π_1 = value of π specified by H_1

   b. $\delta = \gamma\sqrt{N}$

c. $N = \left(\frac{\delta}{\gamma}\right)^2$

d. Conventional values of γ: "small," .10; "medium," 30; "large," .50

3. The significance test of a Pearson r (includes Pearson r with continuous interval scales, r_s, r_{pb}, ϕ)

a. $\gamma = \rho_1$, the value specified by H_1

b. $\delta = \gamma\sqrt{N-1}$

c. $N = \left(\frac{\delta}{\gamma}\right)^2 + 1$

$= \left(\frac{\delta}{\rho_1}\right)^2 + 1$

d. Conventional values of γ: "small," .10; "medium," .30; "large," .50

4. The significance test of the difference between independent means

a. $\gamma = \frac{\theta}{\sigma}$

where θ = value of $\mu_1 - \mu_2$ specified by H_1
σ = population standard deviation (a single value because it is assumed that $\sigma_1 = \sigma_2$)

b. $\delta = \gamma\sqrt{\frac{N}{2}}$

where N is the size of *each* sample (thus $2N$ cases are used in the experiment). If the sample sizes are unequal, use

$$N = \frac{2N_1N_2}{N_1 + N_2}$$

c. $N = 2\left(\frac{\delta}{\gamma}\right)^2$

where N is the size of *each* sample (so $2N$ cases will be needed in all)

d. Conventional values of γ: "small," .20; "medium," .50; "large," .80

Note: The above procedures are approximate, and are applicable for large samples (N at least 25 or 30).

PROBLEMS

1. a. Suppose that the students at Bigbrain U. (Chapter 10, problem 6c) believe that the mean of Bigbrain U. on the SAT is 20 points different (either way) from the national average of 500, and suppose that $\sigma = 100$. Compute the power of the statistical test (using the .05 criterion of significance). What is your evaluation of the research?

b. Now suppose that the students at Bigbrain U. instead believe that they are 50 points different (either way) from the national average but that the other values in part a are not changed. Compute the power of the statistical test; what is your opinion of the research in this case?

c. Explain why power is larger in part b than in part a.

d. Under the conditions of part a, what sample size would be needed to obtain power = .80?

e. Under the conditions of part b, what sample size would be needed to obtain power = .80? Why is this answer much smaller than the one in part d?

2. a. A politician needs 50% or more of the vote to win an election. To find out how his campaign is going, he plans to obtain a random sample of 81 voters and see how many plan to vote for him. He is willing to posit a specific alternative hypothesis of 60% (or 40%) and wishes to use the .01 criterion of significance. Compute the power of the statistical test. How do you evaluate this research plan?

b. Suppose the politician decides to switch to the .05 criterion of significance (but that the other values are not changed). Will this improve the power of the statistical test to a satisfactory level?

c. The politician finally resigns himself to doing more work and obtaining a larger sample. He wishes power = .75. How large a sample does he need (using the .05 criterion)?

3. A personality theorist feels that if two traits are correlated, the correlation should be on the order of .40 (or −.40). She wishes to test the null hypothesis that the correlation between the two traits is .00, using the .05 criterion of significance and a random sample of 65 subjects. Is the power of this statistical test satisfactory?

4. The industrial psychologist in problem 3 of Chapter 11 originally believed that there was a "medium" effect size in the population. Compute the power of this statistical test and briefly evaluate the research plan.

5. Suppose that Smedley finds himself in the following predicament: power = .50, the consequences of a Type II error in his particular experiment are catastrophic, and he cannot possibly get any more subjects. As it happens the consequences of a Type I error in this experiment are not quite so terrible. What should he do?

6. Susanne Trueblood wishes to make up a "Table for the Purpose of Avoiding Poorly Planned Experiments." She feels that an experiment with power less than .70 is not worth running, and she wishes always to have a large enough sample size so that the power is at least this large. For purposes of comparison, she also wishes to know the sample sizes needed for power = .80 and power = .90. She is interested only in the .05 criterion of significance. Complete the table below by filling in the sample sizes.

SAMPLE SIZE AS A FUNCTION OF EFFECT SIZE AND POWER ($\alpha = .05$)

effect size	*power*	*statistical test*			
		mean of 1 population	*proportion of 1 population*	*Pearson r*	*difference, 2 independent means, each N = :*
	.70	________	________	________	________
Small	.80	________	________	________	________
	.90	________	________	________	________
	.70	________	________	________	________
Medium	.80	________	________	________	________
	.90	________	________	________	________
	.70	________	________	________	________
Large	.80	________	________	________	________
	.90	________	________	________	________

3. PEARSON *r*

proposed N	*effect size*	*power, $\alpha = .05$*	*power, $\alpha = .01$*	*N needed for power of .85, if:*	
				$\alpha = .05$	$\alpha = .01$
30	Small	________	________		
100	Small	________	________	________	________
30	Medium	________	________		
100	Medium	________	________	________	________
30	Large	________	________		
100	Large	________	________	________	________

4. DIFFERENCE BETWEEN TWO INDEPENDENT MEANS

proposed N in each sample	*effect size*	*power,* $\alpha = .05$	*power,* $\alpha = .01$	*N in each sample needed for power of .85, if:*	
				$\alpha = .05$	$\alpha = .01$
30	Small	________	________		
100	Small	________	________	________	________
30	Medium	________	________		
100	Medium	________	________	________	________
30	Large	________	________		
100	Large	________	________	________	________

15. One-Way Analysis of Variance

REMINDER

Illustrative example

group:	*1*	*2*	*3*
	10	17	9
	11	23	4
	16	10	2
	13	10	15
Sum	50	60	30

1. Meaning of symbols

SYMBOL	GENERAL MEANING	EXAMPLE VALUE
k	Number of groups (or the last group)	3
N	Total number of observations	12
N_G	Number of observations in group G	-
N_1	Number of observations in group 1	4
N_2	Number of observations in group 2	4
N_3	Number of observations in group 3	4
$\bar{X}$	Grand mean	$140/12 = 11.67$
$\bar{X}_G$	Mean of group G	–
$\bar{X}_1$	Mean of group 1	12.50
$\bar{X}_2$	Mean of group 2	15.00
$\bar{X}_3$	Mean of group 3	7.50
X_G	A score in group G	-

2. Definition and computing formulas

1. *Total sum of squares*

   Definition: $SS_T = \sum (X - \bar{X})^2$

Computing formula: $SS_T = \sum X^2 - \frac{(\sum X)^2}{N}$

Illustrative example:

$$SS_T = (10)^2 + (11)^2 + \cdots + (15)^2 - \frac{(140)^2}{12}$$
$$= 356.67$$

2. *Between-groups sum of squares*

Definition:

$$SS_B = \sum N_G(\bar{X}_G - \bar{X})^2$$

Computing formula:

$$SS_B = \frac{(\sum X_1)^2}{N_1} + \frac{(\sum X_2)^2}{N_2} + \cdots + \frac{(\sum X_k)^2}{N_k} - \frac{(\sum X)^2}{N}$$

Illustrative example:

$$SS_B = \frac{(50)^2}{4} + \frac{(60)^2}{4} + \frac{(30)^2}{4} - \frac{(140)^2}{12}$$
$$= 116.67$$

3. *Within-groups sum of squares*

Definition:

$$SS_W = \sum (X_1 - \bar{X}_1)^2 + \sum (X_2 - \bar{X}_2)^2 + \cdots + \sum (X_k - \bar{X}_k)^2$$

Computing formula: $SS_W = SS_T - SS_B$

Illustrative example: $SS_W = 240$

Illustration of definition formula:

$$SS_W = (10-12.5)^2 + (11-12.5)^2 + (16-12.5)^2 + (13-12.5)^2 + (17-15)^2 + (23-15)^2 + (10-15)^2 + (10-15)^2 + (9-7.5)^2 + (4-7.5)^2 + (2-7.5)^2 + (15-7.5)^2$$
$$= 240$$

3. Steps in one-way analysis of variance

1. Compute SS_T, SS_B, and SS_W.
2. Compute

$$df_B = k - 1$$
$$df_W = N - k$$

3. Compute mean squares (*MS*):

$$MS_B = \frac{SS_B}{df_B}$$

$$MS_W = \frac{SS_W}{df_W}$$

4. Compute

$$F = \frac{MS_B}{MS_W} \qquad df = k - 1, \quad N - k$$

5. Obtain critical value from *F* table and test for significance. If the computed *F* is equal to or greater than the tabled *F*, reject H_0 (that $\mu_1 = \mu_2 = \mu_3$) in favor of H_1 (that H_0 is not true). Otherwise, retain H_0.

6. *If F is statistically significant*, multiple comparisons may be run to determine which of the population means differ from one another. Though this is a complex area, one technique that will usually be appropriate is the Fisher *LSD* (or "protected" *t* test) procedure. For any (or every) pair of groups, compute:

$$t = \frac{\bar{X}_i - \bar{X}_j}{\sqrt{MS_W\left(\frac{1}{N_i} + \frac{1}{N_j}\right)}}, \qquad df = N - k$$

where $\bar{X}_i$ = mean of group *i*
$\bar{X}_j$ = mean of group *j*
MS_W = within-groups mean square
N_i = number of observations in group *i*
N_j = number of observations in group *j*

Do *not* use this procedure if *k*, the number of *groups*, is greater than about 6–8.

4. Measure of strength of relationship

One measure for determining the strength of the relationship between the independent and dependent variables is ε (epsilon):

$$\varepsilon = \sqrt{\frac{df_B(F-1)}{df_B F + df_W}}$$

Epsilon bears the same relationship to F that r_{pb} bears to t (see Chapter 13).

PROBLEMS

1. Below are two hypothetical (separate) sets of data which are somewhat exaggerated to help clarify the procedures underlying analysis of variance. In each case, the experimenter is interested in the number of errors made by rats in a maze as a function of the kind of reward. Group 1 receives 100% water reward; group 2 receives a solution of 50% water and 50% sugar as the reward; and group 3 receives 100% sugar reward.

	EXPERIMENT 1			EXPERIMENT 2		
group:	*1*	*2*	*3*	*1*	*2*	*3*
	1	8	6	4	10	10
	3	7	5	0	0	0
	2	5	3	5	9	8
	1	4	3	0	1	1
	3	6	3	1	10	1
$\sum X_G$	10	30	20	10	30	20
$\bar{X}_G$	2	6	4	2	6	4
		$\bar{X} = 4$			$\bar{X} = 4$	

a. By inspection, in which case would you guess that the difference among groups is more likely to be statistically significant? Why?

b. Carry out the analysis of variance for Experiment 1, using the computing formulas. Are the results statistically significant?

$\sum X^2 =$ ....

$\frac{(\sum X)^2}{N} =$ ....

$SS_T =$ ....

source	*SS*	*df*	*MS*	*F*
between groups	....	....	....	....
within groups (error)	....	....	....	

decision about H_0: ....

2. a. Given that the total sum of squares for the hypothetical University data (page 1) is 2075 and the between-groups sum of squares is 351, perform an analysis of variance for these data. Would you retain or reject the null hypothesis that the four samples come from populations with equal means? (Note that since the sample sizes are very unequal, homogeneity of variance in the populations must be assumed.)

source	*SS*	*df*	*MS*	*F*
between groups	______	______	______	______
within groups (error)	______	______	______	

decision about H_0: ______________________________

b. What advantages does the analysis of variance have over the *t* test of significance when three or more samples are involved?

c. Carry out a multiple comparisons analysis for all possible pairs of Universities in problem 2a, and state your conclusions.

A versus B: $t =$ ____________

A versus C: $t =$ ____________

A versus D: $t =$ ____________

B versus C: $t =$ ____________

B versus D: $t =$ ____________

C versus D: $t =$ ____________

Conclusion: ____________

____

____

____

3. Convert the F values in problems 1 b and 2 a of this chapter to epsilon values. Briefly discuss the implications of each result.

Problem 1 b.

Problem 2 a.

ADDITIONAL COMPUTATIONAL PRACTICE

For each of the two (separate) experiments that follow, compute a one-way analysis of variance. If an experiment yields statistically significant results, compute epsilon and also compute all possible multiple comparisons.

EXPERIMENT 1

group 1	*group 2*	*group 3*
17	15	9
12	11	21
3	4	3
10	26	7
1	18	20
14	23	
5		

source	*SS*	*df*	*MS*	*F*
between groups	----------	----------	----------	----------
within groups (error)	----------	----------	----------	

decision about H_0: --

EXPERIMENT 2

group 1	*group 2*	*group 3*
1	4	26
10	17	21
3	11	15
12	20	9
3	14	23
5	18	
7		

source	*SS*	*df*	*MS*	*F*
between groups	________	________	________	________
within groups (error)	________	________	________	

decision about H_0: ______________________________

16. Two-Way Analysis of Variance

REMINDER Illustrative example (3 × 2 factorial design)

		Factor 1: 1	2	3	row sums
Factor 2	1	5, 4, 4 (sum = 13)	8, 5, 7 (sum = 20)	5, 6, 9 (sum = 20)	53
	2	3, 4, 2 (sum = 9)	7, 5, 6 (sum = 18)	4, 9, 7 (sum = 20)	47
column sums		22	38	40	Total = 100

1. Sums of squares

1. Total sum of squares (SS_T)

   Computing formula: $SS_T = \sum X^2 - \frac{(\sum X)^2}{N}$

   Illustrative example:

   $$SS_T = 5^2 + 4^2 + \cdots + 9^2 + 7^2 - \frac{(100)^2}{18}$$
   $$= 622.0 - 555.56$$
   $$= 66.44$$

2. Between-groups sum of squares (SS_B)

   Computing formula:

   $$SS_B = \frac{(\sum X_1)^2}{N_1} + \frac{(\sum X_2)^2}{N_2} + \cdots + \frac{(\sum X_k)^2}{N_k} - \frac{(\sum X)^2}{N}$$

Illustrative example:

$$SS_B = \frac{(13)^2}{3} + \frac{(20)^2}{3} + \cdots + \frac{(18)^2}{3} + \frac{(20)^2}{3} - \frac{(100)^2}{18}$$
$$= 591.33 - 555.56$$
$$= 35.77$$

3. Within-groups sum of squares (SS_W)

Computing formula: $SS_W = SS_T - SS_B$

Illustrative example:
$$SS_W = 66.44 - 35.77 = 30.67$$

4. Sum of squares for factor 1 (SS_1):

Computing formula:

$$SS_1 = \sum \frac{(\text{Sum of } \textit{each column})^2}{N \textit{ in each column}} - \frac{(\sum X)^2}{N}$$

Illustrative example:

$$SS_1 = \frac{(22)^2}{6} + \frac{(38)^2}{6} + \frac{(40)^2}{6} - \frac{(100)^2}{18}$$
$$= 588.0 - 555.56$$
$$= 32.44$$

5. Sum of squares for factor 2 (SS_2):

Computing formula:
$$SS_2 = \sum \frac{(\text{Sum of } \textit{each row})^2}{N \text{ in each row}} - \frac{(\sum X)^2}{N}$$

Illustrative example:

$$SS_2 = \frac{(53)^2}{9} + \frac{(47)^2}{9} - \frac{(100)^2}{18}$$
$$= 557.56 - 555.56$$
$$= 2.0$$

6. Sum of squares for interaction ($SS_{1 \times 2}$):

Computing formula:
$$SS_{1 \times 2} = SS_B - SS_1 - SS_2$$

Illustrative example:
$$SS_{1 \times 2} = 35.77 - 32.44 - 2.0$$
$$= 1.33$$

2. Degrees of freedom

Total degrees of freedom:

$(df_T) = N - 1$

(where N = total number of observations)

Degrees of freedom within groups:

$(df_W) = N - k$

(where k = number of cells)

Degrees of freedom for factor 1:

(df_1) = one less than the number of levels for factor 1

Degrees of freedom for factor 2:

(df_2) = one less than the number of levels for factor 2

Degrees of freedom for interaction:

$(df_{1\times 2}) = df_1 \times df_2$

3. Mean squares

Mean square within groups: $(MS_W) = \dfrac{SS_W}{df_W}$

Mean square for factor 1: $(MS_1) = \dfrac{SS_1}{df_1}$

Mean square for factor 2: $(MS_2) = \dfrac{SS_2}{df_2}$

Mean square for interaction: $(MS_{1\times 2}) = \dfrac{SS_{1\times 2}}{df_{1\times 2}}$

4. *F* ratios and tests of significance

Effect of factor 1: $F = \dfrac{MS_1}{MS_W}$

Effect of factor 2: $F = \dfrac{MS_2}{MS_W}$

Effect of interaction: $F = \dfrac{MS_{1\times 2}}{MS_W}$

Each computed F value is compared to the critical value from the F table for the degrees of freedom associated with the numerator and denominator *of that test.* If the computed F is less than the table F, H_0 is retained; otherwise, H_0 is rejected in favor of H_1 (the effect is statistically significant).

PROBLEMS

1. Using the data in the illustrative example in the reminder for this chapter, suppose that factor 1 represents severity of mental illness as rated by a clinical psychologist (group 1 = relatively normal, group 2 = mildly neurotic, group 3 = severely neurotic) and factor 2 represents sex (group 1 = males, group 2 = females); cell entries are scores on a written measure of mental adjustment (where high scores indicate maladjustment). Complete the analysis of variance by filling in the table below and carrying out the various tests of significance, and comment briefly on the meaning of the results.

source	*SS*	*df*	*MS*	*F*
severity of illness (columns)	32.44	....	....	....
sex (rows)	2.00	....	....	....
interaction	1.33	....	....	....
error (within groups)	30.67	....	....	

2. An industrial psychologist wishes to determine the effects of satisfaction with pay and satisfaction with job security on overall job satisfaction. He obtains measures of each variable for a total group of 20 employees, and the results are shown below. (Cell entries represent overall job satisfaction, where 7 = very satisfied and 1 = very dissatisfied.) Analyze the data using analysis of variance. What should the psychologist conclude?

		satisfaction with pay	
		high	*low*
satisfaction with job security	*high*	7 7 6 4 6	3 1 2 2 2
	low	1 2 5 2 2	2 1 3 1 1

source	*SS*	*df*	*MS*	*F*
Pay satisfaction (columns)	______	______	______	______
Job security satisfaction (rows)	______	______	______	______
Interaction	______	______	______	______
Error (within groups)	______	______	______	

3. Suppose that a 2 × 2 factorial design is conducted to determine the effects of caffeine and sex on scores on a 20-item English test. The cell means and the various mean squares are given below. Compute the appropriate *F* ratios, test them for statistical significance, and comment briefly on the results.

		caffeine factor	
		caffeine	*placebo*
sex	*males*	$\bar{X} = 17.3$	$\bar{X} = 12.0$
	females	$\bar{X} = 12.3$	$\bar{X} = 16.4$

Mean square for caffeine = 2.40
Mean square for sex = 1.60
Mean square for interaction = 13.84
Mean square within groups = 2.00
Degrees of freedom within groups = 24

ADDITIONAL COMPUTATIONAL PRACTICE

For each of the following (separate) experiments, perform a two-way analysis of variance.

EXPERIMENT 1

		Factor 1		
		1	*2*	*3*
	1	8 17 2	14 10 3	1 7 15
Factor 2	*2*	18 10 19	6 2 2	3 3 7
	3	5 17 6	16 15 9	4 16 1

source	*SS*	*df*	*MS*	*F*
Factor 1	________	________	________	________
Factor 2	________	________	________	________
Interaction	________	________	________	________
Error (within groups)	________	________	________	

EXPERIMENT 2

		Factor 1		
		1	*2*	*3*
Factor 2	*1*	2 3 1	17 14 15	8 10 7
	2	4 2 3	18 16 17	6 5 9
	3	3 2 1	19 15 16	10 6 7

source	*SS*	*df*	*MS*	*F*
Factor 1	________	________	________	________
Factor 2	________	________	________	________
Interaction	________	________	________	________
Error (within groups)	________	________	________	

EXPERIMENT 3

		Factor 1		
		1	*2*	*3*
	1	18 16 17	8 10 7	2 3 1
Factor 2	*2*	4 2 3	19 15 16	6 5 9
	3	10 6 7	3 2 1	17 14 15

source	*SS*	*df*	*MS*	*F*
Factor 1	________	________	________	________
Factor 2	________	________	________	________
Interaction	________	________	________	________
Error (within groups)	________	________	________	

17. Chi Square

REMINDER

Chi square is used with *frequency* data.

$$\chi^2 = \sum \frac{(f_o - f_e)^2}{f_e}$$

where f_o = observed frequency
f_e = expected frequency

1. One-variable problems

$$df = k - 1,$$

where k = number of categories of the variable

Expected frequencies are readily determined from the null hypothesis. For example, if H_0 specifies that subjects in the population are equally divided among the k categories, f_e for each category is equal to N/k (where N = number of subjects in the sample).

2. Two-variable problems; test of association

$$df = (r - 1)(c - 1)$$

where r = number of rows
c = number of columns

For a cell in a given row and column, the expected frequency is equal to

$$f_e = \frac{(\text{row total})(\text{column total})}{N}$$

For a 2×2 table, the following formula for χ^2 is equivalent to determining expected frequencies and using the usual formula, and requires somewhat less work computationally:

A	B
C	D

$$\chi^2 = \frac{N(AD - BC)^2}{(A + B)(C + D)(A + C)(B + D)}$$

3. Measures of strength of association in two-variable tables

2×2 *tables*:

Compute the phi coefficient:

$$\phi = \sqrt{\frac{\chi^2}{N}}$$

ϕ is interpreted as a Pearson r.

Larger tables:

Compute Cramér's ϕ:

$$\text{Cramér's } \phi = \sqrt{\frac{\chi^2}{N(k-1)}}$$

where $k =$ the *smaller* of r (number of rows) or c (number of columns); $k =$ either one if $r = c$

These two measures are statistically significant if χ^2 is statistically significant.

4. Some precautions on the use of χ^2

χ^2 should be used only when the observations are independent—that is, when no observation is related to or dependent upon any other observation.

Do *not* compute χ^2 under any of the following conditions:

a. $df = 1$, and any *expected* frequency is less than 5.

b. $df = 2$, and any *expected* frequency is less than 3.

c. $df = 3$, and more than one *expected* frequency is less than 5 *or* any expected frequency equals 0.

PROBLEMS

1. A coin is flipped 100 times and comes up heads 40 times and tails 60 times. Using χ^2, test the null hypothesis that the coin is "fair."

$\chi^2 =$ ________

$df =$ ________

Decision about H_0: ________

2. An automobile manufacturer observes that in a random sample of 60 adults, 27 prefer blue cars, 19 prefer red cars, and 14 prefer black cars. Using χ^2, test the null hypothesis that the preference for these three colors in the population is equally divided.

$\chi^2 =$ ________

$df =$ ________

Decision about H_0: ________

3. A bond issue is to be put before the voters in a forthcoming election. An opinion poll company obtains a random sample of 200 registered voters and asks them what party they belong to and how they intend to vote on the bond issue. The results are as follows:

		prospective vote on bond issue		
		yes	*no*	*undecided*
political party	*Democratic*	20	30	10
	Republican	30	30	20
	other	10	40	10

Test the null hypothesis that political party and prospective vote on the bond issue are independent. What is your conclusion?

$\chi^2 =$ ____________________

$df =$ ____________________

Decision about H_0: ____________________

4. A psychologist wants to test the hypothesis that college men will do better on a problem-solving task than will college women. He obtains the following results:

		result on problem-solving task	
		succeed	*fail*
sex	*male*	12	18
	female	10	10

Test the null hypothesis that sex and success on the problem-solving task are independent. Does this mean the same as a statement about whether or not the percent success differs between males and females?

$\chi^2 =$ ____________

$df =$ ____________

Decision about H_0: ____________

5. **a.** Convert the result of problem 3 to the appropriate measure of strength of relationship. What is your conclusion as to the strength of the relationship between political party and prospective vote on the bond issue?

b. For a 2×2 table Susanne Trueblood obtains a statistically significant χ^2 of 9.0; $N = 49$. Convert her result to the appropriate correlation coefficient. What should she conclude about the strength of the relationship between the two variables?

c. For a given value of χ^2, are the ϕ and Cramér's ϕ coefficients larger if the total sample size is small or large?

6. Smedley Trueblood performed a study to determine whether or not the attitudes of college students to their influence in college affairs has changed. Two years ago, he asked a sample of 25 students if they had enough of a voice in student affairs; 15 men and 3 women said "no." When the *same* group of 25 students was asked the same question 3 months ago, 1 man and 1 woman said "no." Smedley then set up the following table and computed χ^2:

		attitudes	
		2 years ago	*3 months ago*
sex	*male*	15	1
	female	3	1

Smedley was chagrined to learn from his statistics instructor that this was probably the worst misuse of chi square in the history of statistics. What *three* fundamental rules of χ^2 procedure did Smedley violate?

ADDITIONAL COMPUTATIONAL PRACTICE

At University A, the typical grade distribution is: A, 15%; B, 25%; C, 45%; D, 10%; F, 5%. The grades given by two professors are shown below. For each one (separately), test the null hypothesis that the professor is a "typical" grader, using a one-variable chi square analysis.

PROFESSOR 1

	A	B	C	D	F
f_o	7	13	22	4	6

PROFESSOR 2

	A	B	C	D	F
f_o	13	12	8	3	3

$\chi^2 =$ ____________

Decision about H_0: ____________

$\chi^2 =$ ____________

Decision about H_0: ____________

For each of the two (separate) experiments that follow, test the null hypothesis that the population frequencies are equally divided among the four categories, using a one-variable chi square analysis.

EXPERIMENT 1

	1	*2*	*3*	*4*
f_o	21	14	26	18

EXPERIMENT 2

	1	*2*	*3*	*4*
f_o	28	20	9	22

$\chi^2 =$ ____________

Decision about H_0: ____________

$\chi^2 =$ ____________

Decision about H_0: ____________

For each of the two (separate) experiments that follow, compute a two-variable chi square analysis. If an experiment yields statistically significant results, also compute Cramér's ϕ.

EXPERIMENT 3

	1	2	3
1	20	18	16
2	8	8	30
3	6	19	18

EXPERIMENT 4

	1	2	3
1	11	17	26
2	10	15	21
3	13	13	17

$\chi^2 =$ ______

Decision about H_0: ______

$\chi^2 =$ ______

Decision about H_0: ______

18. Nonparametric and Distribution-Free Methods

REMINDER 1. Basic considerations

The main advantage of nonparametric and distribution-free statistical tests is that they do *not* require the population(s) being sampled to be normally distributed, and therefore are applicable when gross nonnormality is suspected. The primary disadvantage of these methods is that when normality does exist, they are less powerful than the corresponding parametric tests (more likely to lead to a Type II error). A numerical measure of the *power efficiency* of a distribution-free or nonparametric test is

$$\frac{N_p}{N_d} \times 100\%$$

where N_p = sample size required by a parametric test to obtain a specified power for a specified criterion of significance and a specified difference between population means

N_d = sample size required by the corresponding distribution-free or nonparametric test to obtain the same power under the same conditions

The power efficiency of a nonparametric or distribution-free test is almost always less than 100%, and sometimes much less. Thus it is wasteful to use these methods when parametric tests are applicable.

2. The difference between the locations of two independent samples: The rank-sum test

1. Parametric analog: t test for the difference between two independent means (Chapter 11).

   Power efficiency: Approximately 92–95%

2. Computational procedures: Rank *all* scores from 1 (smallest) to N (largest), regardless of which group they are in. In case of ties, follow the usual procedure of assigning the mean of the ranks in question to each of the tied scores. Then compute

$$z = \frac{T_1 - T_E}{\sqrt{\dfrac{N_1 N_2 (N+1)}{12}}}$$

where T_1 = sum of ranks in group 1
$T_E = N_1(N+1)/2$
N_1 = number of observations in group 1
N_2 = number of observations in group 2
N = total number of observations

Do *not* use this procedure if there are fewer than six cases in any group.

3. Measure of strength of relationship: If the value of z computed in the rank-sum test is statistically significant, a measure of the strength of the relationship between group membership and the rank values may be found by computing the *Glass rank biserial correlation* (r_G):

$$r_G = \frac{2(\bar{R}_1 - \bar{R}_2)}{N}$$

where $\bar{R}_1$ = mean of ranks in group 1
$\bar{R}_2$ = mean of ranks in group 2
N = total number of observations

The r_G coefficient is *not* a Pearson r, but does fall between the limits of -1 and $+1$.

3. Differences among the locations of two or more independent samples: The Kruskal–Wallis *H* test

1. Parametric analog: F test of one-way ANOVA (Chapter 15)

   Power efficiency: Approximately 90–95%

2. Computational procedures: Rank *all* scores from 1 (smallest) to N (largest), regardless of which group they are in. In case of ties, follow the usual procedure of assigning the mean of the ranks in question to each of the tied scores. Then compute the sum of squares between groups (SS_B) for the ranks, using either the

usual formula given in Chapter 15 or the following somewhat simpler one that is especially designed for ranked data:

$$SS_B = \frac{T_1^2}{N_1} + \frac{T_2^2}{N_2} + \cdots + \frac{T_k^2}{N_k} - \frac{N(N+1)^2}{4}$$

where T_1 = sum of ranks in group 1
T_2 = sum of ranks in group 2
T_k = sum of ranks in group k
N_1 = number of observations in group 1
N_2 = number of observations in group 2
N_k = number of observations in group k
k = number of groups
N = total number of observations

Once SS_B has been obtained, compute

$$H = \frac{12 SS_B}{N(N+1)}$$

The value of H is then referred to the χ^2 table with $k-1$ degrees of freedom. If H equals or exceeds the tabled value, reject the null hypothesis that all populations have equal locations. Otherwise retain H_0. Do *not* use this procedure if there are only two or three groups *and* any group has fewer than five cases.

3. Multiple comparisons: If (and only if) H is statistically significant, the *protected rank-sum test* may be used to determine which pairs of populations differ significantly in location. First, the two groups being compared must be *reranked*, with all other groups being ignored for purposes of this comparison. Then, the rank-sum test described previously is computed (with N equal to the number of observations *in these two groups*). This test may be performed for any or all pairs of groups.

4. Measure of strength of relationship: If H is statistically significant, a measure of the strength of relationship between group membership and rank on the dependent variable may be obtained by computing epsilon applied to ranks (ε_R):

$$\varepsilon_R = \sqrt{\frac{H-k+1}{N-k}}$$

where k = number of groups
N = total number of observations

4. The difference between the locations of two matched samples: The Wilcoxon test

1. Parametric analog: matched t test (Chapter 11)

   Power efficiency: Approximately 92–95%

2. Computational procedures: First obtain the D score for each subject by subtracting X_2 from X_1. Next, discard any case where $D = 0$ (and reduce N accordingly). Then, *ignoring the sign* of the Ds, rank them from the smallest (rank = 1) to the largest (rank = N). Finally, compute

$$z = \frac{T_1 - T_E}{\sqrt{\frac{(2N+1)T_E}{6}}}$$

   where T_1 = sum of ranks for those D values that are positive (the sum of ranks for those D values that are negative may instead be used)

   $T_E = N(N+1)/4$

   N = number of pairs (excluding cases where $D = 0$)

   Do *not* use this procedure if N is less than about 8.

3. Measure of strength of relationship: If the value of z computed in the Wilcoxon test is statistically significant, a measure of the strength of the relationship between the condition (X_1 versus X_2) and the dependent variable may be found by computing the *matched-pairs rank biserial correlation* (r_C):

$$r_C = \frac{4(T_1 - T_E)}{N(N+1)}$$

   where T_1, T_E, and N have the same meaning as in the Wilcoxon test. The r_C coefficient is *not* a Pearson r, but does fall between the limits of −1 and +1.

5. Rough methods for approximating statistical significance: The median and sign tests

The power efficiency of the median and sign tests is only about 65–70% in most cases, so these methods are used primarily to obtain a quick approximation of the results of more powerful (parametric or rank) tests when samples are large.

The median test is used to compare the locations of two or more independent samples. The first step is to compute the grand median (*G Mdn*) for all N observations. Next organize a $k \times 2$ table, where each sample is divided into two parts: observations falling *above G Mdn*, and observations falling *at or below G Mdn*. Then count and tabulate the frequencies, and compute a two-way chi square (Chapter 17) with $k - 1$ *df*.

The sign test is used to compare the locations of two or more matched samples. First compute D $(= X_1 - X_2)$ for each pair. Next, discard all cases where $D = 0$ and reduce N accordingly. Then compute a one-way chi square (Chapter 17), or use the following (simpler) formula that is designed especially for the sign test:

$$\chi^2 = \frac{(f_p - f_m)^2}{N}, \qquad df = 1$$

where f_p = number of positive Ds
f_m = number of negative Ds
N = total number of pairs (excluding cases where $D = 0$)

PROBLEMS

1. a. Susanne Trueblood argues that since distribution-free tests do not require any assumptions about the distribution in the population, they should always be used in place of parametric tests. Why is this idea incorrect?

b. For a specified difference among population means and for the .05 criterion of significance, a certain distribution-free test requires a sample size of 80 to have power of .80. Under similar conditions, the corresponding parametric test requires a sample size of 60 to have the same power. Compute the power efficiency of the distribution-free test and state in words what this indicates.

2. An operator of a certain machine must turn it off quickly if a danger signal occurs. To test the relative effectiveness of two types of signals, a small group of operators is randomly divided into two groups. Those in group 1 operate machines with a newly designed signal, while those in group 2 use machines with the standard signal. The signal is flashed unexpectedly, and the reaction time of each operator (time taken to turn off the machine after the signal occurs) is measured in seconds. The results are shown below. Use the rank-sum test to determine whether there is a significant difference between the two groups. If there is, compute the appropriate measure of strength of relationship.

group 1	*group 2*
5	3
3	17
4	13
10	2
1	8
3	16
5	6
2	9
7	11
	8

3. a. A psychology professor uses three different methods of instruction in three small classes, with the assignment of students to classes being random, and gives each class the same final examination. The results are shown below. Test the null hypothesis that method of instruction has no effect on examination scores, using the Kruskal–Wallis H test.

group 1	*group 2*	*group 3*
94	60	86
97	97	42
100	96	61
72	57	73
99	93	40
96	90	63
98	92	87
97		65
		67

b. Compute the appropriate measure of the strength of the relationship between method of instruction and examination scores, and comment briefly on the result.

c. Perform the appropriate multiple comparison test for all possible pairs of groups, and comment briefly on the results.

1 versus 2: $z =$ ____________

1 versus 3: $z =$ ____________

2 versus 3: $z =$ ____________

Conclusion: ____________

____

____

4. A different psychology instructor develops a training method that is designed to improve the examination scores of poor students. The performance of a sample of 12 such students on the posttest (X_1) following training, and the pretest (X_2) prior to training, is shown below. Test the null hypothesis that the training method has no effect, using the Wilcoxon test. If the result is statistically significant, compute the appropriate measure of strength of relationship.

S	X_1	X_2
1	77	68
2	64	64
3	56	52
4	56	57
5	71	69
6	54	58
7	76	70
8	57	60
9	63	61
10	68	61
11	71	73
12	66	66

5. a. Why are the median and sign tests used primarily as rough approximations only?

b. Since the main advantage of the median test is computational ease, it is usually used only with large samples. As an illustration, however, compute the median test for the data in problem 3a of this chapter. Would a similar conclusion about significance be reached?

c. Similarly, the sign test is usually used only with large samples. As an illustration, however, compute the sign test for the data in problem 4 of this chapter. Would a similar conclusion about significance be reached?

ADDITIONAL COMPUTATIONAL PRACTICE

For each of the two (separate) experiments that follow, perform the rank-sum test. If an experiment yields statistically significant results, also compute r_G.

EXPERIMENT 1		EXPERIMENT 2	
group 1	*group 2*	*group 1*	*group 2*
43	80	96	43
53	45	78	54
57	62	41	59
41	83	60	70
62	46	53	45
54	87	54	83
63	56	63	57
46	70	46	80
59	75	80	50
50	78	56	87
54	89	75	57
57	96	46	62
57		57	
80		62	
60		89	

$z =$ ________

Decision about H_0: ________

$z =$ ________

Decision about H_0: ________

For each of the two (separate) experiments that follow, perform the Kruskal–Wallis H test. If an experiment yields statistically significant results, compute ε_R and also perform all possible multiple comparisons.

EXPERIMENT 3			EXPERIMENT 4		
group 1	*group 2*	*group 3*	*group 1*	*group 2*	*group 3*
36	43	26	36	33	43
8	33	28	33	6	36
33	6	19	22	26	28
14	11	36	14	19	38
26	22	46	28	28	43
43	24	9	11	8	46
28	28	31	26	30	42
35	36	38	31	9	48
42	40		35	24	
30			36		
48			40		

$H =$ ____________

Decision about H_0: ____________

$H =$ ____________

Decision about H_0: ____________

For each of the two (separate) experiments that follow, perform the Wilcoxon test. If an experiment yields statistically significant results, also compute r_C.

EXPERIMENT 5

S	X_1	X_2
1	12	9
2	13	8
3	23	16
4	17	21
5	19	14
6	20	9
7	15	17
8	14	9
9	22	14
10	18	13
11	25	10
12	11	10
13	20	15
14	15	15
15	10	11

$z =$ ____________

Decision about H_0: ____________

EXPERIMENT 6

S	X_1	X_2
1	15	15
2	25	8
3	9	16
4	13	9
5	11	23
6	19	14
7	10	21
8	12	18
9	15	17
10	14	17
11	9	14
12	11	15
13	20	13
14	10	10
15	20	22

$z =$ ____________

Decision about H_0: ____________

Review Section II. Review of Inferential Statistics

For each of the following problems, select the appropriate procedure from the answer column below, compute the answer, and comment briefly on the meaning of the results.

Answer Column

a. "Normal curve" problem (raw scores)

b. Standard error of the mean

c. Standard error of a proportion

d. Confidence intervals

e. t test for independent samples (and r_{pb})

f. t test for matched samples

g. Pearson correlation coefficient (r)

h. Rank-order correlation coefficient

i. Linear regression

j. One-way analysis of variance (and epsilon)

k. Two-way analysis of variance

l. Chi-square (and phi; Cramér's phi)

m. power (may be used in combination with other procedures)

n. Rank-sum test (and r_G)

o. Kruskal–Wallis H test (and the protected rank-sum test; epsilon applied to ranks)

p. Wilcoxon test (and r_C)

PROBLEMS

1. A psychologist wishes to test the theory that people in high-level jobs and people in low-level jobs differ in job satisfaction. She obtains a random sample of 51 executives and 51 assembly-line workers and gives each group a measure of job satisfaction. The results are as follows:

	executives	*assembly-line workers*
$\bar{X}$	80.65	70.15
s	12.00	14.00

What should the psychologist decide?

2. A dean wishes to determine if liking for statistics is influenced by the instructor of the course. A total of 500 students who have had different statistics instructors are asked whether they liked statistics, disliked statistics, or were neutral. The results are as follows:

	instructor A	*instructor B*	*instructor C*
Liked statistics	70	70	60
Disliked statistics	10	140	50
Neutral	20	40	40

What should the dean decide?

3. Suppose that the mean annual sale for all statistics workbooks in the United States is 5000 copies and the standard deviation is 1000.

a. If a workbook must sell at least 4200 copies per year for the publisher to break even, what percent of the workbooks in the country are either breaking even or making a profit?

b. If the Workbook Best Seller List includes those workbooks ranking in the top 15% of the United States in sales, what is the minimum number of copies of this workbook that must be sold per year for it to make the best seller list?

4. A professor believes that "poor students" (those with a scholastic average of C− or less) at his University are more diligent than the typical poor student and spend much more time studying. He knows that the national average studying time for "poor students" is six hours per week. His research plan is to obtain data from 25 poor students at his university and use $\alpha = .05$.

a. If the population effect size is posited as "medium," is the probability of obtaining statistical significance sufficient to proceed with the experiment? If not, what change would you recommend?

b. If, regardless of the results of part a, the professor proceeds with the experiment as is and finds that the mean hours studying per week for a sample of 25 "poor students" is 8.0 and the standard deviation is 6.0, what should he conclude?

5. A psychologist theorizes that tranquilizers will help the test performance of highly anxious people but may harm the performance of calm people by making them too sleepy to perform well. Ten "anxious" subjects (as determined by a clinician's evaluation) are given a tranquilizer; ten "anxious" subjects receive a placebo; ten calm subjects receive a tranquilizer; and ten calm subjects receive a placebo. (Of course, subjects do *not* know what they are given.) All subjects then take a test. The experimenter computes the following results: "Tranquilizer" sum of squares = 2.1; "anxiety" sum of squares = 1.8; interaction sum of squares = 12.3; within-group sum of squares = 54.0. Complete the statistical analysis and state your conclusions.

6. An industrial psychologist wishes to determine an index of worker's pay controlled for tenure (length of time on the job). She would like to use pay as a criterion of job performance, but she knows that a ten-year worker will be making more money than a one-year worker because of greater seniority and that it is therefore not fair to compare the two without taking length of time on the job into account. For a sample of 40 employees, she obtains the following data:

	pay	*tenure*
$\bar{X}$	\$8000	8 years
σ	\$2000	4 years

$$r_{\text{pay, tenure}} = .50$$

The tenure and actual pay of each member of a five-person department is shown below. For each worker, determine if he or she is above average, average, or below average with regard to pay when tenure is taken into account. (*Hint:* Compute the predicted pay for each worker.)

worker	*tenure*	*actual annual pay* (\$)
1	2 years	8000
2	10 years	8000
3	5 years	7250
4	4 years	7500
5	4 years	6500

7. A common belief is that job satisfaction and job productivity are correlated. A psychologist who wishes to test the null hypothesis that the correlation is zero against the alternative that the correlation is .40 (or −.40) administers a job satisfaction scale to 50 workers and also obtains a measure of productivity. He calculates the results shown below. Complete the analysis; what should he decide *and* how confident should he be about his decision?

$S =$ satisfaction measure, $P =$ productivity measure

$\sum S = 100$, $\sum S^2 = 218$, $\sum P = 80$, $\sum P^2 = 136$, $\sum SP = 161$

8. A company advertises that 80% of American housewives use its product. A federal committee investigating the accuracy of this claim obtains a sample of 81 housewives and finds that 75% use the company's product. What conclusion should the committee reach?

9. Subjects are randomly assigned to four conditions of an experiment on visual-motor displays. In each condition, the same calibrated dial is used and the subject must report the reading on the dial as quickly and as accurately as possible. The hypothesis under study is that subjects who are motivated will make more correct responses (out of a total of 10 trials) on the dial-reading task. Four kinds of motivation are used: a simple instruction to "do your best"; punishment (the subject receives a mild shock for each error); reward (the subject receives a cash award for each correct response); and group competition (subjects compete in a group and the one with the fewest errors wins a prize). There are three subjects in each condition, and the results are as follows:

Group:

1 ("Do your best")	*2 (punishment)*	*3 (reward)*	*4 (competition)*
4	6	7	5
3	5	8	3
3	9	10	7

Does the kind of motivation affect the number of correct responses?

10. Two professors each evaluate an incoming group of eight new graduate students by placing them in order of apparent ability. Professor A decides that Smith is best, Johnson is second, Jones is third, and Richardson is fourth; he cannot decide among Ellis, Morgan, and Bryant for the next position, but does feel that Green is worst. Professor B rates Jones best, and Ellis second. She cannot decide among Smith, Johnson, and Richardson for the next position, but does feel that Bryant is worse than these three students, Morgan is next to worst, and Green is worst. Do the professors agree substantially?

11. A psychologist hypothesizes that husbands are more dominant than wives. He obtains dominance scores for seven couples; the results are shown below. What should the psychologist decide?

couple	*husband's score*	*wife's score*
1	27	20
2	24	26
3	12	42
4	34	30
5	38	28
6	39	32
7	50	44

12. An instructor obtains the following scores on this review chapter from his class of nine students:

9 8 9 13 6 8 6 5 8

He wishes to know all reasonably likely values of the mean of the population of statistics students who complete this review chapter.

13. A psychologist wishes to know if men and women differ in problem-solving ability. She obtains a sample of 25 men and a sample of 25 women and gives each person a number of problems to solve. The results are shown below.

	men	*women*
Mean number of problems solved	16.2	14.2
s^2	14.0	11.0

a. Prior to the experiment, the psychologist posits the effect size in the population as "medium." Test the results for statistical significance; what is your decision concerning H_0?

b. If the experiment is to be repeated using new samples, what sample sizes would you recommend?

14. An experimenter wishes to test the data shown below for differences in location among the four groups, and strongly suspects that the distributions in the corresponding populations are grossly non-normal. Compute the appropriate distribution-free test. Should the null hypothesis of equal population locations be retained?

group			
1	*2*	*3*	*4*
9	8	7	5
7	9	5	3
10	6	8	8
9	8		4
4			

15. An experimenter wishes to perform a test of significance for the matched data below, and strongly suspects that the distributions in the corresponding populations are grossly nonnormal. Compute the appropriate distribution-free test. Should the null hypothesis of equal population locations be retained?

S	X_1	X_2
1	4	6
2	9	6
3	11	9
4	12	10
5	13	10
6	10	8
7	14	4
8	3	4
9	13	6
10	5	3
11	14	6
12	7	7
13	13	8
14	8	12
15	15	7

Answers to Selected Problems

CHAPTER 1

1. (a) $\sum X + \sum Y$ (b) $\sum G + \sum P^2$

3. (b) $\sum X = 55; \sum X^2 = 685; (\sum X)^2 = 3025$

Additional Computational Practice Problems:

	set 1	*set 3*
N	5	14
$\sum X$	7	1176
$\sum Y$	11	1129
$\sum X^2$	15	100,288
$\sum Y^2$	39	93,343
$(\sum X)^2$	49	1,382,976
$(\sum Y)^2$	121	1,274,641
$\sum XY$	23	96,426
$\sum X \sum Y$	585	1,327,704
$\sum(X+Y)$	18	2305
$\sum(X-Y)$	−4	47
Multiply each X by 3.2	22.4	3763.2
Subtract 7 from each Y	−24	1031
Add 1.8 to each X	16	1201.2
Divide each Y by 4	2.75	282.25

CHAPTER 2

1.

score	*univ. A*		*univ. B*		*univ. C*		*univ. D*	
	f	*cf*	*f*	*cf*	*f*	*cf*	*f*	*cf*
20	1	50	0	50				
19	0	49	1	50				
18	2	49	0	49				
17	4	47	3	49			1	5
16	4	43	2	46			0	4
15	5	39	0	44			0	4
14	2	34	4	44			1	4
13	7	32	4	40			0	3
12	7	25	8	36			0	3
11	5	18	5	28	1	10	0	3
10	4	13	6	23	0	9	1	3
9	4	9	4	17	2	9	0	2
8	1	5	3	13	0	7	1	2
7	0	4	2	10	1	7	0	1
6	1	4	2	8	1	6	1	1
5	0	3	2	6	2	5		
4	1	3	0	4	2	3		
3	0	2	2	4	0	1		
2	1	2	1	2	0	1		
1	0	1	1	1	0	1		
0	1	1	0	0	1	1		

2.

class interval	*univ. A*		*univ. B*	
	f	*cf*	*f*	*cf*
20	1	50	0	50
18–19	2	49	1	50
16–17	8	47	5	49
14–15	7	39	4	44
12–13	14	32	12	40
10–11	9	18	11	28
8–9	5	9	7	17
6–7	1	4	4	10
4–5	1	3	2	6
2–3	1	2	3	4
0–1	1	1	1	1

Note: An interval size of 3 is also acceptable.

CHAPTER 3

1.

	$L\%$	LRL	h	$I\%$	PR
(a)	8%	4.5	1	4%	10%

2. (a) $.25 \times 50 = 12.5$th case, which is in the 8 "interval." $LRL = 7.5$; $SFB = 10$; $f = 3$; $h = 1$; $Score_{.25} = 8.3$

3.

	$L\%$	LRL	h	$I\%$	PR
(a)	88%	15.5	2	10%	90.5%

4. (a) $.30 \times 50 = 15$th case, which is in the 8–9 interval. $LRL = 7.5$; $SFB = 10$; $f = 7$; $h = 2$; $Score_{.30} = 8.9$

Additional Computational Practice Problems:

Table 1

Score	*H%*	*I%*	*L%*	*PR*
99	0%	2.70%	97.30%	98.65%
97	2.70%	2.70%	94.59%	95.94%
94	10.81%	8.11%	81.08%	85.14%
91	51.35%	16.22%	32.43%	40.54%
88	86.49%	8.11%	5.41%	9.46%
85	97.30%	2.70%	0%	1.35%

Table 2

p	pN	LRL	SFB	f	$Score_p$
.99	36.63	98.5	36	1	99.13
.80	29.60	92.5	26	4	93.40
.60	22.20	91.5	18	8	92.02
.50	18.50	91.5	18	8	91.56
.25	9.25	89.5	7	5	89.95
.20	7.40	89.5	7	5	89.58
.01	0.37	84.5	0	1	84.87

CHAPTER 4

1. (a) University C: $\bar{X} = 6.00$; University D: $\bar{X} = 11.00$.

2. $\sum fX = 60$; $\bar{X} = 6.00$

3. $LRL = 11.5$, $Mdn = 12.5$

6.
$$\begin{aligned}
11 - 6 &= +5\\
9 - 6 &= +3\\
9 - 6 &= +3\\
7 - 6 &= +1\\
6 - 6 &= 0\\
5 - 6 &= -1\\
5 - 6 &= -1\\
4 - 6 &= -2\\
4 - 6 &= -2\\
0 - 6 &= -6\\
\hline
\textstyle\sum &= 0
\end{aligned}$$

7. (b) Use the mean. $\sum X = 63$; $\bar{X} = 4.20$ min

Additional Computational Practice Problems:

data set 1: $\bar{X} = 19/9 = 2.11$
data set 3: $\bar{X} = 152/29 = 5.24$;
$Mdn = 4.5 + (14.5 - 9)/7 = 5.29$

Chapter 1 problems:

	$\bar{X}$	
data set 1	1.40	2.20
data set 3	84.00	80.64

CHAPTER 5

3. (c) $s^2 = 10.00$; $s = 3.16$; $\sigma^2 = 9.00$; $\sigma = 3.00$
(d) $s^2 = 20.00$; $s = 4.47$; $\sigma^2 = 16.00$; $\sigma = 4.00$
(e) The same results as in (d) should be obtained

4. (c) 11 (d) 11

Additional Computational Practice Problems:

	σ	s
data set 1:	1.79	1.90
data set 3:	1.87	1.90

Chapter 1 problems:

	X	Y
data set 1	$\sigma = 1.02$	$\sigma = 1.72$
	$s = 1.14$	$s = 1.92$

data set 3	$\sigma = 10.36$	$\sigma = 12.81$
	$s = 10.76$	$s = 13.29$

CHAPTER 6

1.

	$\bar{X}$	σ	σ^2
(a)	14.8	4	16
(c)	25.6	12.8	163.84

4.

	Z score	*T score*
(a)	+1.19	61.9
(c)	+0.12	51.2
(e)	0.00	50.0

Additional Computational Practice Problems:

Data set 3, Chapter 1: $\bar{X} = 84.00$, $\sigma_X = 10.36$
$\bar{Y} = 80.64$, $\sigma_Y = 12.81$

S	z_X	z_Y
1	+1.25	+0.65
2	−1.54	−1.85
3	+0.10	+0.50
4	−0.97	−0.36
5	+0.77	+1.28
6	+0.77	−0.13
7	+1.54	+0.81
8	−2.03	−2.39
9	+0.10	+0.34
10	+0.29	+0.26
11	−0.29	+0.81
12	+0.87	+0.81
13	−0.68	−0.44
14	−0.19	−0.28

CHAPTER 7

1. University A:

	12.5	
10.4		15.2
0		20

CHAPTER 8

1. (a) 2/3 (d) 5 to 1

2. (a) 1/13 (c) 1/52 × 1/52

3. (b) 5/13 × 2/51

5. (a) 1/2 (b) 1/128 or .008

Additional Computational Practice Problems:

A. (1) 1/100, or .01 (2) .01 (3) .02 (4) .05 (5) .10
(6) .50 (7) .085 (8) .31

C. (1) .25 (2) .36 (3) .81 (4) 1.00 (5) .01
(6) 0

CHAPTER 9

1. (a) $z = -0.90$, % = 31.59; Answer = 18.41%
(c) $z = -2.25$, % = 48.78; $z = -1.25$, % = 39.44; Answer = 9.34%
(e) $z = +1.65$; Answer = 665

Additional Computational Practice Problems:

Table 1

percent of the population:	*z value(s)*	*percent(s) from table*	*answer*
below 55	−0.78	28.23	21.77%
below 75	+1.01	34.38	84.38%
above 70	+0.56	21.23	28.77%
between μ and 85.7	+1.96	47.50	47.50%
between 60 and 70	−0.33, +0.56	12.93, 21.23	34.16%

Table 2

raw score required to be in top:	*z value*	*answer*
½%	+2.58	92.6
2½%	+1.96	85.7
10%	+1.28	78.0
50%	0	63.7
80%	−0.84	54.3
95%	−1.65	45.2
99%	−2.34	37.5

CHAPTER 10

5. (b) (1) This is technically not random, but no great harm would come of treating this as a random sample.

(3) Not random if students choose their own seats; for example, better students may sit near the front.

6. (a) $s_{\bar{X}} = .56$; $t = 2.96$; reject H_0

(c) No; $s_{\bar{X}} = 20$; $t = 1.50$; retain H_0

7. Yes; $\sigma_p = .05$; $z = 2.00$; reject H_0.

9. $s_{\bar{X}} = .57$

(a) $ts_{\bar{X}} = 1.53$; interval $= 8.99–12.05$

(c) $ts_{\bar{X}} = 0.95$; interval $= 9.57–11.47$

Additional Computational Practice Problems:

	data set 1, chapter 4	*data set 3, chapter 4*
mean	2.11	5.24
s	1.90	1.90
N	9	29
$s_{\bar{X}}$	0.63	0.35
t	−0.62	−2.15
decision about H_0	retain	reject
95% confidence interval	0.65 to 3.57	4.52 to 5.96
99% confidence interval	−0.01 to 4.23	4.27 to 6.21

p	σ_p	z	*decision about H_0*	*decision about profit*
29%	0.028	3.18	reject	profit

CHAPTER 11

2. $s_A^2 = 15.74$; $s_B^2 = 15.97$; $s_{\text{pooled}}^2 = 15.86$; $df = 98$; $t = 1.82/.80 = 2.28$; $p < .05$; reject H_0; University A wins.

5. $\sum D = 10$; $\sum D^2 = 52$; $s_D^2 = 4.67$; $t = 1.45$; retain H_0; there is not sufficient reason to conclude that the two procedures differ.

6. Column 1,
reading down: 154 196 263 25 31 42 11 13 17
Column 3,
reading down: 616 785 1051 70 88 118 26 32 43

Additional Computational Practice Problems:

	power, $\alpha = .05$	*power,* $\alpha = .01$	*N,* $\alpha = .05$	*N,* $\alpha = .01$
Table 2	.08	.02		
	.17	.06	900	1304
	.36	.16		
	.85	.66	100	145
	.77	.55		
	.99	.99	36	53
Table 4	.13	.04		
	.29	.12	450	652
	.48	.25		
	.94	.82	72	105
	.87	.70		
	.99	.99	29	41

CHAPTER 15

1. (b) $\sum X^2 = 302$; $(\sum X)^2/N = 240$; $SS_T = 62$; $SS_B = 40$; $SS_W = 22$; $MS_B = 20$; $MS_W = 1.83$; $F_{2,12} = 10.93$; $p < .01$; Reject H_0; conclude that kind of reward affects performance.

(e) F was statistically significant only in Experiment 1, so multiple comparisons should be run only for this experiment.
Groups 1 and 2: $t = 4/.856 = 4.67$, $p < .01$.
Groups 1 and 3: $t = 2/.856 = 2.34$, $p < .05$.
Groups 2 and 3: $t = 2/.856 = 2.34$, $p < 0.5$.
Conclusion: The means of the three populations from which these three groups (samples) were drawn all differ from one another.

2. (a) $SS_B = 351$; $SS_W = 1724$; $df_W = 111$; $df_B = 3$; $MS_W = 15.52$; $MS_B = 117$; $F_{3,111} = 7.5$; $p < .01$. Reject H_0 and conclude that students at different Universities perform differently on the psychology quiz.

3. In Problem 2(a), epsilon = .38, so there is a moderate relationship between University and performance on the psychology quiz.

Additional Computational Practice Problems:

experiment 1

source	*SS*	*df*	*MS*	*F*
between groups	172.81	2	86.40	1.62
within groups	797.69	15	53.18	

decision: retain H_0; there is not sufficient reason to reject the null hypothesis that the means of the three populations are equal.

CHAPTER 16

1.

source	*SS*	*df*	*MS*	*F*
severity of illness	32.44	2	16.22	6.34
sex	2.00	1	2.00	0.78
interaction	1.33	2	0.67	0.26
error	30.67	12	2.56	

The *F* value for severity of illness is statistically significant, so there is a relationship between the psychologist's diagnosis and scores on the written test. (It would appear that the test distinguishes between relative normals and neurotics but not between mild and severe neurotics, but additional statistical analysis is needed to test this possibility; H_1 states only that the three groups do not come from populations with equal means.) There is not sufficient reason to conclude that sex is related to test scores, or that there is an interaction effect. (Note: Means and *N*s are obtained by referring to the raw data in the reminder.)

3. Only the interaction is statistically significant ($F = 6.92$ for 1 and 24 degrees of freedom). Thus, there is not sufficient reason to conclude that caffeine on the average, or sex on the average, is related to test scores; but there is sufficient reason to conclude that there is a *joint* effect of the two variables on test scores—apparently, that caffeine is beneficial for men and harmful for women.

Additional Computational Practice Problems:

experiment 1

source	*SS*	*df*	*MS*	*F*
factor 1	112.96	2	56.48	1.73 (*NS*)
factor 2	20.51	2	10.26	0.31 (*NS*)
interaction	233.71	4	58.43	1.79 (*NS*)
error	588.00	18	32.67	

experiment 2

source	*SS*	*df*	*MS*	*F*
factor 1	900.96	2	450.48	186.92 ($p < .001$)
factor 2	0.52	2	0.26	0.11 (*NS*)
interaction	10.37	4	2.59	1.07 (*NS*)
error	43.33	18	2.41	

CHAPTER 17

1. Chi square $= 4.0$; $df = 1$; $p < .05$; reject H_0; conclude that coin is "loaded."

4. Chi square $= .49$; $df = 1$; retain H_0; there is not sufficient reason to conclude that sex and success on this task are related. This is equivalent to saying that the percent success is not significantly different between the sexes.

5. (a) Cramér's phi $= .18$; the relationship is weak.

Additional Computational Practice Problems:

Professor 1: Chi square $= 4.89$, $df = 4$. Retain H_0; there is not sufficient reason to reject the hypothesis that this professor is a "typical" grader.

Experiment 2: Chi square $= 9.56$, $df = 3$, $p < .05$. Reject H_0.

Experiment 3: Chi square $= 18.84$, $df = 4$, $p < .001$. Reject H_0. Cramér's $\phi = .26$.

CHAPTER 18

2. $N_1 = 9$, $N_2 = 10$, $N = 19$, $\sum R = (19)(20)/2 = 190$, $T_1 = 63.5$, and $T_2 = 126.5$. Using group 1, $T_E = 90$, $\sigma_T = 12.25$, and $z = -26.5/12.25 = -2.16$; $p < .05$, and the new signal is superior. (Using group 2, $T_E = 100$, $z = +2.16$, and the conclusion is identical.) $r_G = -.59$, indicating a strong relationship in the sample with those in group 1 reacting more quickly (having smaller reaction times).

4. After deleting the two cases where $D = 0$ (cases 2 and 12), $N = 10$ and $\sum R = (10)(11)/2 = 55$. T_1 (for the positive Ds) $= 39.5$, T_2 (for the negative Ds) $= 15.5$, $T_E = 27.5$, and $z = 12/9.81 = 1.22$, which is not statistically significant. Thus there is not sufficient reason to believe that the training method has an effect on examination scores.

5. (b) $N = 24$, so G $Mdn = 88.5$ (halfway between the 12th and 13th case). The resulting $k \times 2$ table is

	1	*2*	*3*
above *G Mdn*	7	5	0
at or below *G Mdn*	1	2	9

$\chi^2 = 14.79$ and $p < .001$ for 2 *df*, so a similar result is obtained. (c) $\chi^2 = (6 - 4)^2/10 = .4$, which is not statistically significant, so a similar result is obtained.

Additional Computational Practice Problems:

Experiment 2: $T_1 = 213.5$, $T_2 = 164.5$, $\sum R = 378$, T_E using group 1 $= 210$, $z = 3.5/20.49 = 0.17$, retain H_0.

Experiment 4: $T_1 = 150$, $T_2 = 72$, $T_3 = 184$, $\sum R = 406$, $SS_B = 966.45$, $H = 14.28$, $df = 2$, $p < .001$, reject H_0. $\varepsilon_R = .70$. Multiple comparisons: 1 versus 2, not quite statistically significant ($T_1 = 140.5$, $T_2 = 69.5$, $z = 25/13.16 = 1.90$), 1 versus 3 is statistically significant ($T_1 = 75.5$, $T_2 = 114.5$, $z = 34.5/12.11 = 2.85$, $p < .01$), and 2 versus 3 is statistically significant ($T_1 = 47.5$, $T_2 = 105.5$, $z = 33.5/10.39 = 3.22$, $p < .01$).

Experiment 6: $N = 13$, $T_1 = 33.5$, $T_2 = 57.5$, $\sum R = 91$, $T_E = 45.5$, $z = 12/14.31 = 0.84$, retain H_0.

A 8
B 9
C 0
D 1
E 2
F 3
G 4
H 5
I 6
J 7